Introduction to Cartography

Introduction to Cartography

Caroline Rivera

www.callistoreference.com

Callisto Reference,
118-35 Queens Blvd., Suite 400,
Forest Hills, NY 11375, USA

Visit us on the World Wide Web at:
www.callistoreference.com

ISBN: 978-1-64116-574-7 (Hardback)

Cataloging-in-Publication Data

Introduction to cartography / Caroline Rivera.
p. cm.
Includes bibliographical references and index.
ISBN 978-1-64116-574-7
1. Cartography. 2. Maps. 3. Map projection. I. Rivera, Caroline.
GA101.5 .I58 2022
526--dc23

Table of Contents

Preface

This book aims to help a broader range of students by exploring a wide variety of significant topics related to this discipline. It will help students in achieving a higher level of understanding of the subject and excel in their respective fields. This book would not have been possible without the unwavered support of my senior professors who took out the time to provide me feedback and help me with the process. I would also like to thank my family for their patience and support.

The study and practice of making maps is referred to as cartography. It is mainly concerned with the modeling of reality such that effective communication can take place regarding spatial information. This discipline can be broadly divided into two categories, namely, general cartography and thematic cartography. General cartography caters to a general audience and thus can contain a variety of different features. Thematic cartography focuses on using specific geographic themes which are aimed at a select target audience. Modern cartography uses computer software such as CAD, GIS and specialized illustration software for making maps. Some of the symbols which are used in maps are legend, compass rose, bar scale and title. The topics included in this book on cartography are of utmost significance and bound to provide incredible insights to readers. It aims to shed light on some of the unexplored aspects of this field. This book will provide comprehensive knowledge to the readers.

A brief overview of the book contents is provided below:

Chapter - Introduction

Cartography is the science and technique of graphically representing a geographical area on a flat surface, in the form of maps or charts. This chapter will briefly explain the different types of cartography such as celestial cartography, planetary cartography, critical cartography, terrain cartography, etc.

Chapter - Elements and Concepts of Cartography

Some of the major elements and concepts related to cartography are symbology, map legend, contour line, scale bars, compass rose, map projection, cartographic labeling, chorography, cartographic generalization, photogrammetry, etc. The topics elaborated in this chapter will help in gaining a better perspective about these elements and concepts.

Chapter - Maps and its Types

Maps are the illustrations or the diagrammatical depictions of an area of land, sea or space, which portray physical features of cities, roads, etc. Some of the different types of maps are pictorial maps, nautical charts, world maps, reference maps, topographical maps, etc. This chapter has been carefully written to provide an easy understanding of these types of maps.

Chapter - Web Cartography and Web-enabled GIS

Web cartography, also known as web mapping, is the process of using maps delivered by GIS on the World Wide Web. Some of the well-known web maps are Yahoo maps, Bing maps and Google maps. This chapter closely examines these key web maps to provide an extensive understanding of the web cartography and web-enabled GIS.

Chapter - Interdisciplinary Fields Involving Cartography

There are numerous fields which involve the application of cartography. A few of them are geodesy, geomatics and topography. Geodesy deals with the study of the shape and area of the Earth, geomatics focuses on the collection, storage and analysis of geographic data or geographic information and topography studies the shape and features of land surfaces. The diverse applications of cartography in these fields have been thoroughly discussed in this chapter.

Caroline Rivera

1

Introduction

Cartography is the science and technique of graphically representing a geographical area on a flat surface, in the form of maps or charts. This chapter will briefly explain the different types of cartography such as celestial cartography, planetary cartography, critical cartography, terrain cartography, etc.

CARTOGRAPHY

Cartography or mapmaking is the study and practice of making representations of the Earth on a flat surface. The discipline of cartography combines science, aesthetics, and technical ability to create a balanced and readable representation that is capable of communicating information effectively and quickly.

A celestial map from the seventeenth century.

Cartography, however mechanized it becomes, remains both a science and an art. The aesthetics of any given map will always be a critical component essential to the conveyance of information. A map must provide accuracy and in the best of solutions, an inventive presentation of data or analysis of data, but always in a form that is readily comprehensible and inviting to the reader. A map is both more, and less, than simply geographical or physical space. And it is always a result of artistic and technical judgments, creating something both useful and, occasionally, beautiful.

One problem in creating maps is the simple reality that the surface of the Earth, a curved surface in three-dimensional space, must be represented in two dimensions as a flat surface. This necessarily entails some degree of distortion, which can be dealt with by utilizing projections that minimize distortion in certain areas. Furthermore, the Earth is not a regular sphere, but its shape is instead known as a geoid, which is a highly irregular but exactly knowable and calculable shape.

Maps of all scales have traditionally been drawn and made by hand, but the use of computers has revolutionized cartography. Most commercial-quality maps are now made with software that falls into one of three main types: CAD, GIS, and specialized illustration software.

Functioning as tools, maps communicate spatial information by making it visible. Spatial information is acquired from measurement of space and can be stored in a database, from which it can be extracted for a variety of purposes. Current trends in this field are moving away from analog methods of mapmaking and toward the creation of increasingly dynamic, interactive maps that can be manipulated digitally.

Cartographic representation involves the use of symbols and lines to illustrate geographic phenomena. This can aid in visualizing space in an abstract and portable format. The cartographic process rests on the premise that the world is measurable and that we can make reliable representations or models of that reality.

Maps have been a large part of the human story for a long time (perhaps 8,000 years - nobody knows exactly, but longer than written words). They were known to have existed in societies of Europe, the Middle East, China, India, and others.

The earliest known map to date is a wall painting of the ancient Turkish city of Çatal Hüyük which has been dated to the late seventh millennium B.C.E. Other known maps of the ancient world include the Minoan "House of the Admiral" wall painting from c. 1600 B.C.E. showing a seaside community in an oblique perspective, and an engraved map of the holy Babylonian city of Nippur, from the Kassite period (fourteenth – twelfth centuries B.C.E.). The ancient Greeks and Romans created maps beginning with Anaximander in the sixth century B.C.E. In ancient China, although geographical literature spans back to the fifth century B.C.E., the drawing of true geographical maps was not begun in earnest until the first half of the Han Dynasty, with the works of Prince Liu An.

Mappa mundi is the general term used to describe Medieval European maps of the world. Approximately 1,100 mappae mundi are known to have survived from the Middle Ages. Of these, some 900 are found illustrating manuscripts and the remainder exist as stand-alone documents.

In the Age of Exploration from the fifteenth century to the seventeenth century, cartographers copied earlier maps (some of which had been passed down for centuries) and drew their based on explorers' observations and new surveying techniques. The invention of the magnetic compass, telescope, and sextant increased accuracy.

Due to the sheer physical difficulties inherent in cartography, map-makers frequently lifted material from earlier works without giving credit to the original cartographer. For example, one of the most famous early maps of North America is unofficially known as the Beaver Map. This map is an exact reproduction of a 1698 work by Nicolas de Fer. De Fer in turn had copied images that were first printed in books by Louis Hennepin, and François Du Creux. By the 1700s, map-makers started to give credit to the original engraver by printing the phrase "After the original cartographer" on the work.

Not all maps were drawn on paper. Well researched examples include the navigational stick charts of the Marshall Islanders, interwoven sticks arranged to depict distances across seas, wave fronts, and elevations of islands. Native Alaskans carved intricate sculptures that recreated coastlines and elevations in a portable, and quite accurate, three dimensional form.

Technological Changes

In cartography, new technology has been incorporated into the production of the maps of new generations of mapmakers and map users. The first maps were manually constructed with brushes and parchment, were varied in quality and of limited distribution. The advent of magnetic devices, like the compass and, much later, magnetic storage devices, led to the creation of far more accurate maps and the ability to store and manipulate those maps digitally.

Advances in mechanical devices such as the printing press, quadrant, and vernier calipers allowed for the mass production of maps and the ability to make accurate reproductions from more accurate data. Optical technology, such as the telescope, sextant, and other devices that use telescopes, allowed for accurate surveying of land and gave the mapmakers and navigators the ability to find their latitude by measuring angles to the North Star at night or the sun at noon.

Advances in photochemical technology, such as the lithographic and photochemical processes, have allowed for the creation of maps that are finely detailed, do not distort in shape, and resist moisture and wear. These advances eliminated the need for engraving, further shortening the time it takes to make and reproduce maps.

In the late twentieth century and early twenty-first century, advances in electronic technology led to another revolution in cartography. Specifically, computer hardware devices such as computer screens, plotters, printers, scanners (remote and document), and analytic stereo plotters along with visualization, image processing, spatial analysis and database software, have democratized and greatly expanded the making of maps. The ability to superimpose spatially located variables onto existing maps created new uses for maps and new industries to explore and exploit these potentials.

Map Types

The field of cartography can be divided into two broad categories: general cartography and thematic cartography. General cartography involves those maps that are constructed for a general audience and thus contain a variety of features, like topographic maps. Topographic maps depict natural and built features of a place, with relief and elevation shown by drawn contours or shading techniques. These relatively general maps exhibit many reference and location systems and often are produced in a series. For example, United States Geological Survey (USGS) has produced a full series of 1:24,000 scale topographic maps; Canada has the same, at 1:50,000 scale. The government of the UK produces 1:63,360 (1 inch to 1 mile) "Ordnance Survey" maps of the entire UK and a range of correlated larger- and smaller-scale maps of great detail.

Thematic cartography involves maps of specific geographic themes oriented toward specific audiences. Examples might be a dot map showing corn production in Indiana or a shaded area map of Ohio counties divided into numerical choropleth classes. As the volume of geographic data has exploded over the last century, thematic cartography has become increasingly useful and necessary to interpret spatial cultural and social data. Epidemiological data are represented on specialized maps, a particularly useful way to illustrate exposure patterns, or occurrence. Most applied cartography could be well be described as thematic mapping. Points of view can be represented thematically as well, and the user of a given map must be informed of the objectives of the cartographer in order to judge the value of the presentation.

Map Design

Arthur H. Robinson, an American cartographer influential in thematic cartography, stated that a poorly designed map "will be a cartographic failure." He also declared that "map design is perhaps the most complex" aspect of cartography. Robinson codified the mapmaker's understanding that a map must be designed with consideration of the audience and its needs foremost. A well designed map would address each of these basic elements:

- Ease of use, with respect to the intended audience, both physically and cognitively; accuracy, meaning a minimum amount of distortion or errors;
- Strong relationship between the object and the map, meaning that the translation of physical space to a different medium should be readily recognizable;
- Appropriate labeling and symbol use;
- Legibility and clarity - very important points.

From the very beginning of mapmaking, maps "have been made for some particular purpose or set of purposes." The intent of the map should be illustrated in a manner in which the 'percipient' acknowledges its purpose in a timely fashion. The term percipient refers to the person receiving information and was used by Robinson. The figure-ground principle refers to this notion of engaging the user by clear presentation, leaving no confusion concerning the purpose of the map. Clear presentation enhances the user's experience and keeps his attention. If the user is unable to identify what is being demonstrated, the map may be useless.

Making a meaningful map is the ultimate goal. MacEachren explains that a well designed map "is convincing because it implies authenticity". A thoughtfully designed, interesting map engages a reader. Information richness or a map that is multivariate will show relationships within the map. Showing several variables allows comparison, adding to the meaningfulness of the map. This also generates hypotheses, stimulates ideas, and perhaps, further research.

In order to convey the message of the map, the creator must design it in a manner that will facilitate the overall understanding of its purpose. The title of a map may provide the "needed link" necessary for communicating that message, but the overall design of the map fosters the manner in which the reader interprets it.

Naming Conventions

Most maps use text to label places and for such things as a map title, legend, and other information. Maps are typically created in specific languages, though names of places often differ among languages. So a map made in English may use the name Germany for that country, where a German map would use Deutschland, and a French map Allemagne. A word that describes a place using a non-native terminology or language is referred to as an exonym.

In some cases, the 'correct' name is unclear. For example, the nation of Burma officially changed its name to Myanmar, but many nations do not recognize the ruling junta and continue to use Burma. Sometimes an official name change is resisted in other languages and the older name may remain in common use. Examples include the use of Saigon for Ho Chi Minh City, Bangkok for Krung Thep, and Ivory Coast for Côte d'Ivoire.

Difficulties arise when transliteration or transcription between writing systems is required. National names tend to have well established names in other languages and writing systems, such as Russia for Росси́я, but for many place names a system of transliteration or transcription is required. In transliteration the symbols of one language are represented by symbols in another. For example, the Cyrillic letter Р is traditionally written as R in the Latin alphabet. Systems exist for transliteration of Arabic, but the results may vary. For example, the Yemeni city of Mocha is written variously in English as Mocha, Al Mukha, al-Mukhā, Mocca, and Moka. Transliteration systems are based on relating written symbols to one another, while transcription is the attempt to spell the phonetic sounds of one language in another. Chinese writing is transformed into the Latin alphabet through the Pinyin phonetic transcription systems, for example. Other systems were used in the past, such as Wade-Giles, resulting in the city being spelled Beijing on newer English maps and Peking on older ones.

Further difficulties arise when countries, especially former colonies, do not have a strong national geographic naming standard. In such cases cartographers may have to choose between various phonetic spellings of local names versus older imposed, sometimes resented, colonial names. Some countries have multiple official languages, resulting in multiple official place names. For example, the capital of Belgium is both Brussels and Bruxelles. In Canada, English and French are official languages and places are named in both languages. British Columbia is also officially named la Colombie-Britannique. English maps rarely show the French names outside Quebec, which itself is spelled Québec in French.

The study of placenames is called toponymy, while that of the origin and historical usage of placenames as words is etymology.

Map Symbolization

The quality of a map's design affects its reader's ability to comprehend and extract information from the map. Cartographic symbolization has been developed in an effort to portray the world accurately and effectively convey information to the map reader. A legend explains the pictorial language of the map, or its symbolization. The title indicates the region the map portrays or the map's intent; the map image portrays the region and so on. Although every map element serves some purpose, convention dictates inclusion of only certain elements while others are considered optional. A menu of map elements includes the neatline (border), compass rose or north arrow, overview map, scale bar, projection, and information about the map sources, accuracy, and publication.

When examining a landscape, scale can be intuited from trees, houses, and cars. Not so with a map. Thus a simple thing as a north arrow can be crucial; the top of a map does not necessarily indicate north.

Color is equally important. How the cartographer uses color to display the data can greatly affect the clarity or intent of the map. Different intensities of hue portray the cartographer's various objectives. Computers can display up to 16 million distinct colors at a time even though the human eye can distinguish only a minimum number of these. This allows for a multitude of color options for even for the most complex maps. Moreover, computers can easily hatch patterns in colors to give even more options. This can be very useful when symbolizing data in categories like quintile and equal interval classifications.

Quantitative symbols give a visual measure of the relative size/importance/number that a symbol represents. There are two major classes of symbols used for portraying quantitative properties on a map: Proportional symbols change their visual weight according to a quantitative property. These are appropriate for extensive statistics. Choropleth maps portray data collection areas (such as counties, or census tracts) with color. Using color this way, the darkness and intensity (or value) of the color is evaluated by the eye as a measure of intensity or concentration.

Map Generalization

A good map is a compromise between portraying the items of interest (or themes) in the right place for the map scale used, and the need to annotate that item with text or a symbol, taking up space on the map medium and very likely causing some other item of interest to be displaced. The cartographer is thus constantly making judgments about what to include, what to leave out, and what to show in a slightly incorrect place because of the demands of the annotation. This issue assumes more importance as the scale of the map gets smaller (i.e., the map shows a larger area), because relatively, the annotation on the map takes up more space on the ground. A good example from the late 1980s was the British Government Ordnance Survey's first digital maps, where the absolute positions of major roads shown at scales of 1:1250 and 1:2500 were sometimes a scale distance of hundreds of meters away from ground truth, when shown on digital maps at scales of 1:250000 and 1:625000, because of the overriding need to annotate the features.

CARTOGRAPHIC PROCESS

Today, maps can be produced easily through a wide range of online tools by anyone with access to the Internet. Maps used in most activities (from urban planning, through geological exploration or environmental management, to trip planning and navigation), however, are still typically produced by professionals with expertise in mapping or in the phenomena being depicted on the maps. The academic and professional field that focuses on mapping is called "cartography." Cartography has been defined by the International Cartographic Association as "the discipline dealing with the conception, production, dissemination and study of maps." One useful conceptualization of cartography is as a process that links map makers, map users, the environment mapped, and the map itself.

The cartographic process is a cycle that begins with a real or imagined environment. As map makers collect data from the environment (through technology and remote sensing), they use their perception to detect patterns and subsequently prepare the data for map creation (i.e., they think about the data and its patterns as well as how to best visualize them on a map). Next, the map maker uses the data and attempts to signify it visually on a map (encoding), applying generalization, symbolization, and production methods that will (hopefully) lead to a depiction that can be interpreted by the map user in the way the map maker intended (its purpose). Next, the map user reads, analyzes, and interprets the map by decoding the symbols and recognizing patterns. Finally, users make decisions and take action based upon what they find in the map. Through their provision of a viewpoint on the world, maps influence our spatial behavior and spatial preferences and shape how we view the environment.

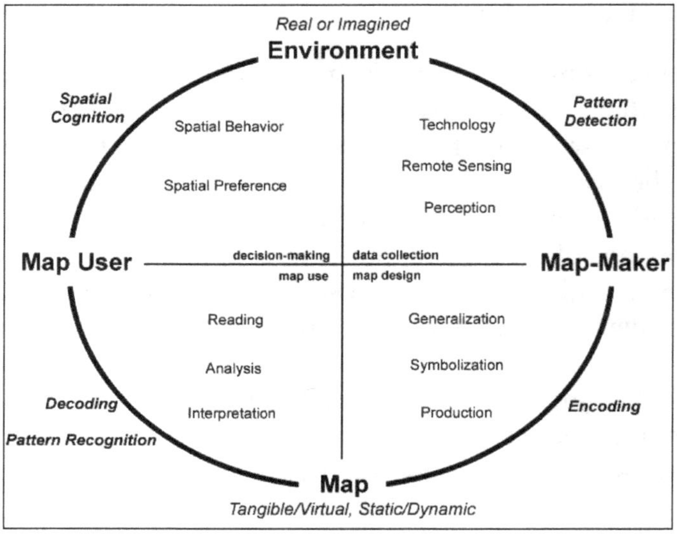

The Cartographic Process.

In the cartographic process as outlined above, the fundamental component in generating a map to depict the environment is itself a process – the process of map abstraction.

Map Abstraction

It has become possible to map the world on the head of a pin, or even a smaller space, but, most details get left out. Even to achieve a screen-sized map of the world on your computer, map abstraction is fundamental to representing entities in a legible manner. The process of map abstraction includes at least five major (interdependent) steps: (a) Selection, (b) Classification, (c) Simplification, (d) Exaggeration, and (e) Symbolization.

Selection

Depending on a map's purpose, cartographers (map makers) select what information to include and what information to leave out. The cartographer must answer four questions: Where? When? What? Why? As an example, a cartographer can create a map of San Diego (where) showing current (when) traffic patterns (what) so that an ambulance can take the fastest route to an emergency (why).

The map in Figure shows how a cartographer selected specific highways to include along with a few other features; these other features include a very generalized representation of the terrain, a few major rivers and lakes, and an indication of the area included in each of several communities (in pastel colors). The objective is to help drivers pick efficient routes by depicting the highways and whether traffic is moving quickly (green) or stalled (red). Other information is kept to a minimum and visually pushed to the background; that extra information is included to provide context for the primary focus (the highways and traffic on them).

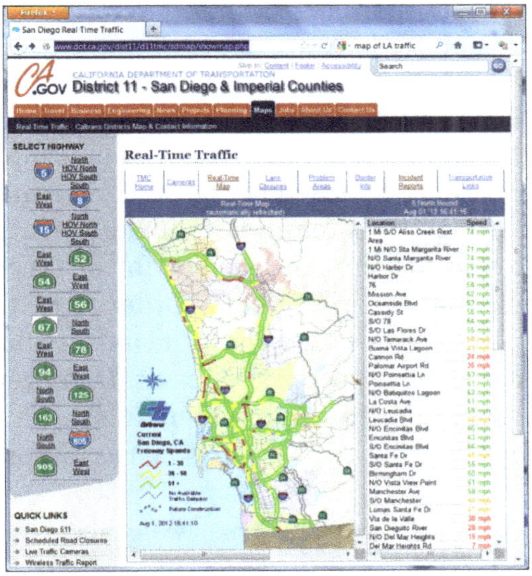

Screenshot of San Diego Real-Time Traffic Application; to try out the map.

Classification

Classification is the grouping of things into categories, or classes. By grouping attributes into a few discernible classes, new visual patterns in the data can emerge and the map becomes more legible. In the example above, the highways are classified into those without traffic detectors (gray) and those with traffic detectors (in color) and furthermore, within the latter, into slow (red), intermediate (yellow), and fast (green) travel conditions. Map readers must consider about classification, the example below shows one dataset for the rate of prostate cancer by county in Pennsylvania mapped using a different number of classes. As you can see, different patterns emerge depending upon how many classes the cartographer chooses to visualize. One must be critical when looking at maps because changing the map classification can change what appears to be true.

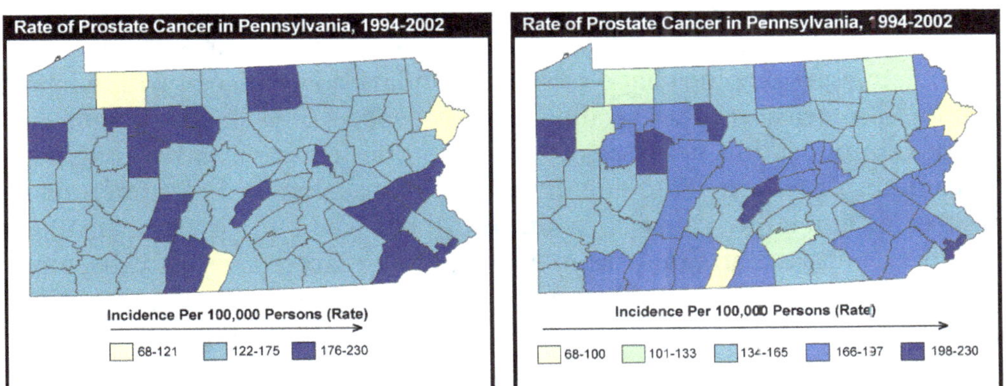

Incidence rate of prostate cancer per 100,000 persons per county in Pennsylvania, visualized using three classes (left) and five classes (right).

Simplification

Cartographers also need to simplify the features on a map beyond the tasks of feature type selection and feature classification in order to make a map more intelligible. This includes choosing to delete, smooth, typify, and aggregate entities within feature types. In the process of deleting entities, imagine creating a map of cities for the United States. As illustrated in Figure, attempting to include every city in the U.S. would render the map illegible. Map makers must delete, for instance, cities below a certain population (as done in the map on the right) in order to better serve the purpose of the map. In this case, if the purpose was to show the most populous cities, a fixed population threshold produces a very appropriate result. If, however, the purpose was to show the most important cities in the region, then an arbitrary population threshold does not work since, for example, Salt Lake City is just as important to Utah as Phoenix is to Arizona.

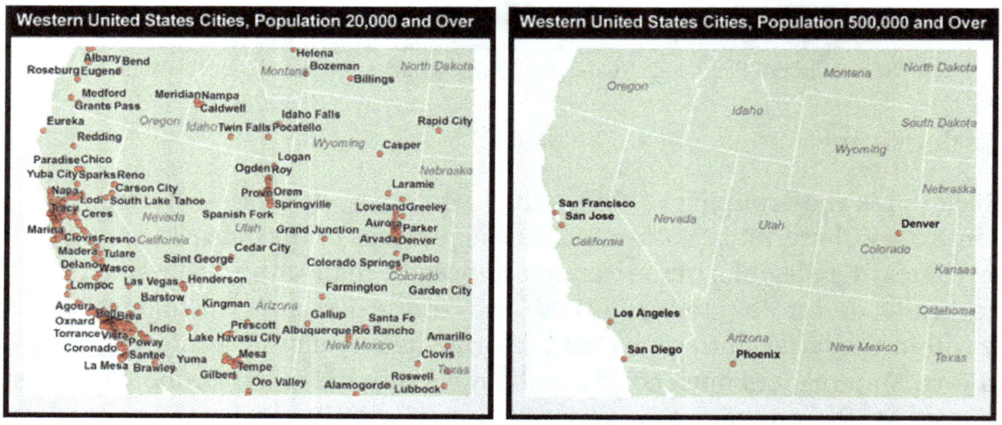

Simplification of cities in the western United States by deleting cities with populations below 500,000.

Smoothing is the act of eliminating unnecessary elements in the geometry of features, such as the superfluous details of a nation's shoreline that can only be seen at a larger, zoomed-in regional scale. Typification depicts just the most typical components of the mapped feature. The visibility map above is a good example of typification in which the actual geographic shape of state boundaries is replaced with what might be considered a caricature that retains only key aspects of each state's shape. Going beyond the simplification processes that act on one feature at a time, aggregation combines multiple features into one. Imagine a river composed of numerous meandering streams at a large scale (i.e., zoomed in), but when moving to a smaller scale (i.e., zooming out), the streams are merged into one larger river as it becomes impossible to maintain the detail. If you visit Google Maps and zoom in to Harrisburg, Pennsylvania, you will find the Susquehanna River flowing through the middle of the capital. As you zoom out to a smaller scale, you will view the various smaller streams of the Susquehanna begin to collapse into a single blue line as the details of the river aggregate.

Exaggeration

Deliberate exaggeration of map features is often performed in order to allow certain features to be seen. For instance, on a standard paper highway map of Pennsylvania (the fold-up kind you might have in the glove box of your car, thus about 3 feet across when unfolded), interstate highways are printed at roughly 0.035 inches in width. That sounds pretty small, right? But, if the width of the printed road relative to the map width was the same as the width of the actual highway relative to the width of Pennsylvania, it would mean that the Interstate was nearly 2000 feet wide. This is a typical case of exaggeration to create an abstraction that is useful for travel.

Symbolization

In the final process of creating a map, the cartographer symbolizes the selected features on a map. These features can be symbolized in visually realistic ways, such as a river depicted by a winding blue line. But many depictions are much more abstract, such as a circle or star representing a city. Map symbols are constructed from more primitive "graphic variables, the elements that make up symbols. Below, we provide a brief overview of these core graphic variables; then we focus on how color in particular is used (or should be used).

Graphic Variables

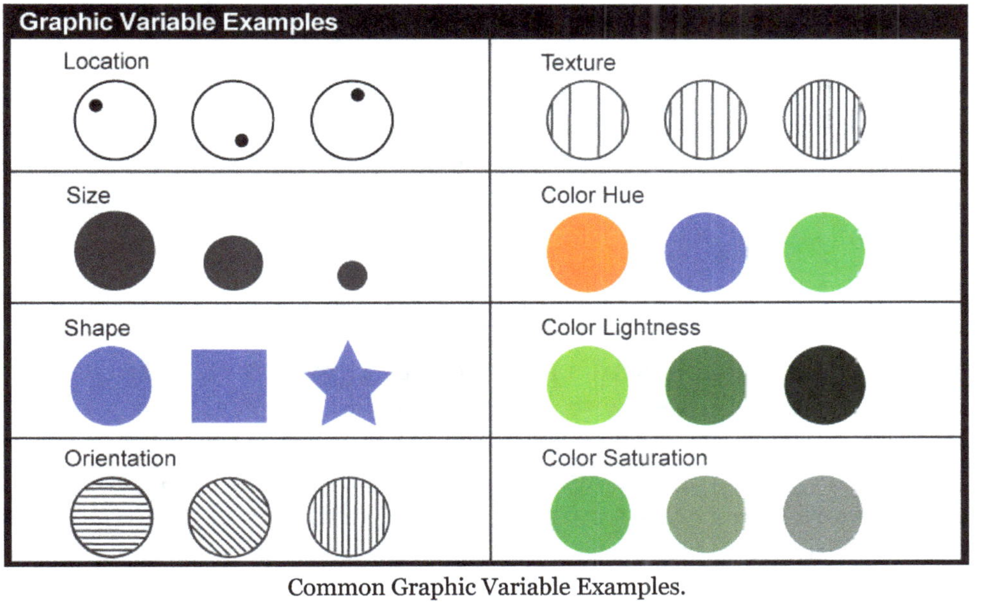

Common Graphic Variable Examples.

Given the large variety of maps that exist, it might be surprising to learn that the visual appearance of all maps starts from a very small set of display primitives from which all those variations can be constructed. We call these primitives graphic variables because each represents a "graphic" (visible) feature of a map symbol that can be "varied." While different cartographers have identified a slightly different set of primitives, most agree that there are somewhere between 7 and 12 of them from which all maps symbolization can be constructed. The most commonly cited primitives that can be varied for map symbols are: location, size, shape, orientation, texture, and three components of color – color hue (red, green, blue, etc.), color lightness (how light or dark the color is), color saturation (how pure the color hue is). By convention, each of these "graphic variables" is used to represent particular categories of data variation.

Color Schemes

Three of the graphic variables are components of color. Color is particularly important for map symbolization today since so many maps are seen online where color is always available and nearly always used. While most maps you will see use color to depict data (as well as in aesthetic ways), many maps do not use color in the most logical ways in relation to the data being depicted. Well-designed maps use variations in the three color variables in ways that reflect the kinds of variations in the underlying data they represent. Recognizing the latter is particularly important so that you are not misled by maps you encounter.

To help cartographers (and others) select good colors for maps, Dr. Cynthia Brewer and Dr. Mark Harrower developed Color Brewer, a web app designed to help users pick colors based on data type, number of data classes, and mode of map presentation (i.e., printing, photocopying). The color schemes have been tested with users who have color deficiency (about 8% of the population; difficulty distinguishing red from green is the most common). The web app allows users to interact with a map template by changing colors, background, borders, and terrain. There are three main color scheme forms a user can choose from: sequential, diverging, and categorical. Each is appropriate for specific kinds of data as detailed below.

Sequential color schemes should be employed when data is arranged from a low to a high data value (e.g., data for mean annual income by county in Pennsylvania). This sequential scheme aligns colors from light (depicting low data values) to dark (depicting high data values) in a step-wise sequence. Sequential schemes can rely on only color lightness as shown below at left or may add some color hue variation to enhance differences in categories will retaining the clear visual ordering as shown at right. As an example, Figure uses a 4-class purple sequential scheme to depict Avian Influenza, with a focus on Eurasia.

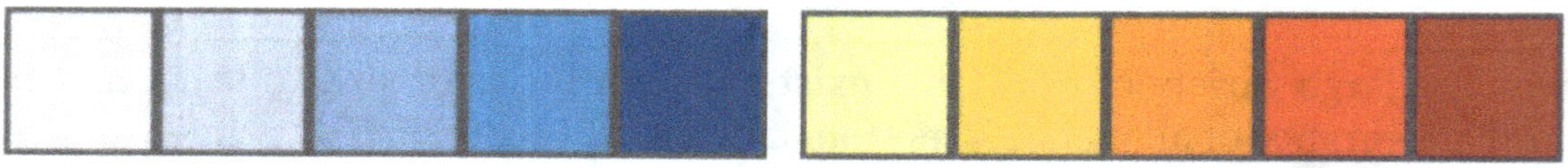

Screenshot of a single hue sequential color scheme for 5 classes (left) and a multi-hue sequential color scheme for 5 classes (right).

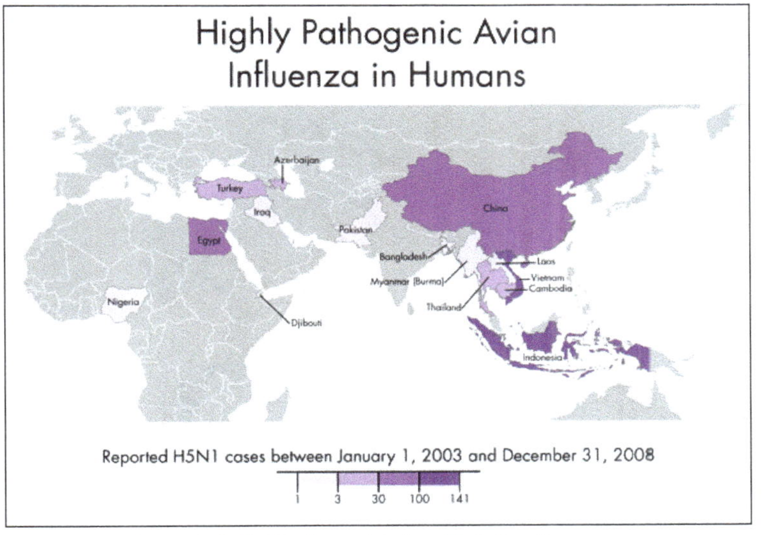

Reported H5N1 Cases (Avian Flu) Per Country.

Diverging color schemes highlight an important midrange or critical value of ordered data as well as the maximum and minimum data values. Two contrasting dark hues converge in color lightness at the critical value. This is the scheme used for the population change map in figure below in which the critical dividing point is zero change.

Diverging color scheme for 5 classes.

Unlike the ordered data mentioned in the previous color schemes, qualitative color schemes are used to present categorical data, or data belonging to different categories. Different hues visually separate each of the different classes, or categories.

DESIGN PRINCIPLES FOR CARTOGRAPHY

Cartographers apply many design principles when compiling their maps and constructing page layouts. Five of the main design principles are legibility, visual contrast, figure-ground, hierarchical organization, and balance. Together these form a system for seeing and understanding the relative importance of the content in the map and on the page. Without these, map-based communication will fail. Together visual contrast and legibility provide the basis for seeing the contents on the map. Figure-ground, hierarchical organization and balance lead the map reader through the contents to determine the importance of things and ultimately find patterns.

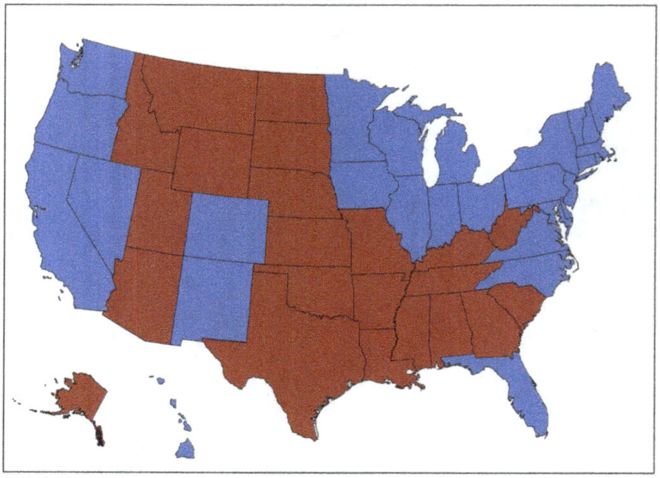

Visual Contrast

Visual contrast which relates to how map features and page elements contrast with each other and their background. To understand this principle at work, consider your inability to see well in a dark environment. Your eyes are not receiving much reflected light so there is little visual contrast between features and you cannot easily distinguish objects from one another or from their surroundings. Add more light and you are now able to contrast features from the background. This concept of visual contrast also applies in cartography. A well-designed map with a high degree of visual contrast can result in a crisp, clean, sharp-looking map. The higher the contrast between features, the more something will stand out, usually the feature that is darker or brighter. Conversely, a map that has low visual contrast can be used to promote a more subtle impression. Features that have less contrast will appear to belong together.

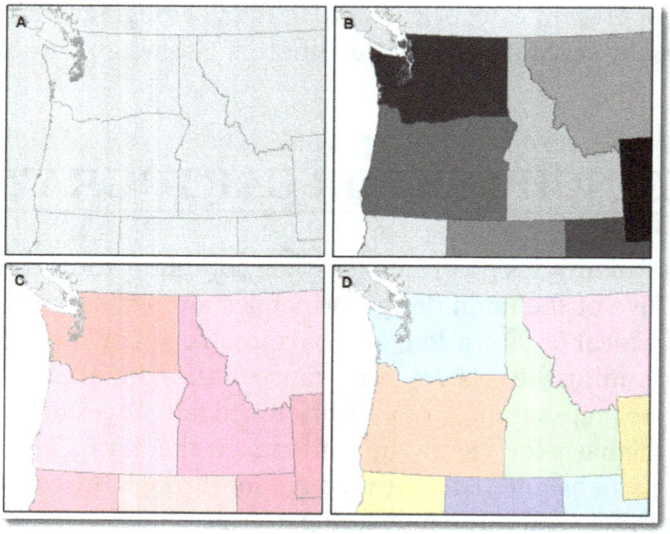

In figure, when there is no variation in visual contrast (A), the map reader has a hard time distinguishing features from the background. For quantitative distributions (B), there must be enough contrast between tones for the reader to distinguish unique classes. For qualitative distributions (C),using variations of a single color hue (e.g., red) does not provide as much contrast as using a variety of hues (e.g., red, green, blue, etc).

Legibility

Legibility is "the ability to be seen and understood". Many people work to make their map contents and page elements easily seen, but it is also important that they can be understood. Legibility depends on good decision-making for selecting symbols that are familiar and choosing appropriate sizes so that the results are effortlessly seen and easily understood. Geometric symbols are easier to read at smaller sizes; more complex symbols require larger amounts of space to be legible.

Visual contrast and legibility are the basis for seeing. In addition to being able to distinguish features from one another and the background, the features need to be large enough to be seen and to be understood in order for your mind to decipher what you eyes are detecting. Visual contrast and legibility can also be used to promote the other design principles: figure-ground, hierarchical organization, and balance.

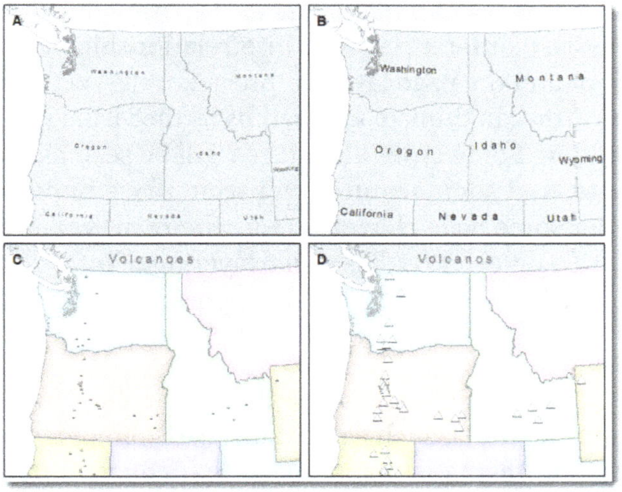

Text and symbols (A and C) that are too small cannot be seen. Once able to be seen (B and D), they must also be understood.

Figure-ground

Figure-ground organization is the spontaneous separation of the figure in the foreground from an "amorphous" background. Cartographers use this design principle to help their map readers find the area of the map or page to focus on. There are many to promote figure-ground organization, such as adding detail to the map or using a white wash, a drop shadow, or feathering.

Using closed forms (A), a white wash (B), a drop shadow (C), or feathering (D) will promote figure-ground organization on your map.

Hierarchical Organization

One of the major objectives in map making is to "separate meaningful characteristics and to portray likenesses, differences, and interrelationships". The internal graphic structuring of the map (and the page layout more generally) is fundamental to helping people read your map. You can think of a hierarchy as the visual separation of your map into layers of information. Some types of features will be seen as more important than other kinds of features, and some features will seem more important than other features of the same type. Some page elements (e.g., the map) will seem more important than others (e.g., the title or legend). This visual layering of information within the map and on the page helps readers focus on what is important and enables them to identify patterns.

Hierarchical organization on reference maps (those that show the location of a variety of physical and cultural features, such as terrain, roads, boundaries, and settlements) works differently than on thematic maps (maps that concentrate on the distribution of a single attribute or the relationship among several attributes). For reference maps, many of the features should be no more important than one another and so, visually, they should lie on essentially the same visual plane. In reference maps, hierarchy is usually more subtle and the map reader brings elements to the forefront by focusing attention on them. For thematic maps, the theme is more important than the base that provides geographic context.

Balance

Balance involves the organization of the map and other elements on the page. A well-balanced map page results in an impression of equilibrium and harmony. We can also use balance in different ways to promote "edginess" or "tension" or create an impression that is more "organic". Balance results from two primary factors, visual weight and visual direction. If you imagine that the center of your map page is balancing on a fulcrum, the factors that will "tip" the map in a particular direction include the relative location, shape, size, and subject matter of elements on the page.

In figure, which of the top six maps seems most balanced? It should appear that (F) has visual equilibrium, usually achieved by placing the central figure slightly above center on the page. However, the addition of page elements, such as the title and legend, will modify the visual impression, so all content on the page should be evaluated together to judge balance.

Together these five design principles have a significant impact on your map. How they are used will either draw the attention of your map readers or potentially repel them. Giving careful thought to the design of your maps using these principles will help you to assure that your maps are ones people will want to look at.

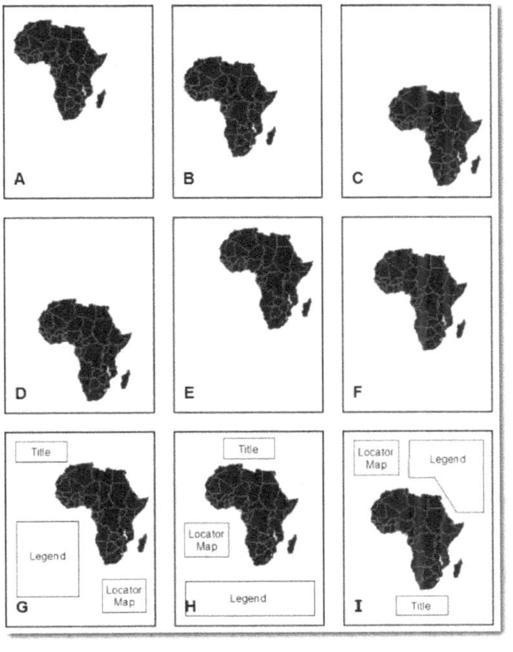

CELESTIAL CARTOGRAPHY

Celestial cartography, also known as uranography or star cartography, is the area of astronomy that produces star maps. Most are aware of cartographers who made early and more recent maps of Earth. Both types of cartography were difficult in early history; we were not certain of the exact sizes and shapes of land masses, and there are many, many stars to chart correctly. Many early Earth and star maps were very artistic in nature and truly took not only a scientific mind, but the skills of an amazing artist.

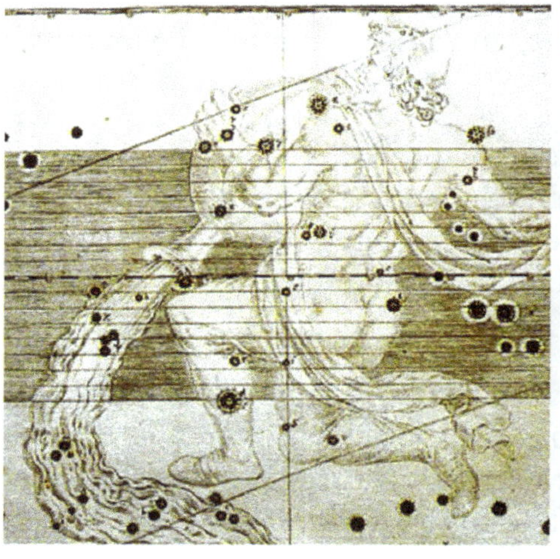

All around the world ancient cultures celebrated gods, heroes, and mythical beasts by giving names to groups of stars. About 3000 years ago in Assyria ancient people recorded information regarding the constellations on stone tablets. Between 1300 and 1600 B.C. Babylonian and Greek astronomers added their constellations and today 48 of them are still in use. The Babylonian systems of math predicated star formation and constellation alignment and were later adopted by the Greeks in the 4th century.

Boundaries

The ancient astronomers did not have defined boundaries within the constellations and and some stars overlapped into other constellations. For example the star Alpheratz found a home in the constellation of Pegasus and Andromeda. Stars of the southern sky, not visible to ancient astronomers of northern latitudes, were not grouped into constellations.

Early Celestial Cartography

With the advent of copper engraved plates for printing illustrations, Cartography greatly improved. Copper engravings provided several advantages over woodcuts. Copper allowed the artist a chance at producing finer details and the plates did not have the durability issues found in woodcuts. Overall quality of the maps improved.

Newer and more accurate star catalogs became available and the invention of the telescope enabled further exploration of the solar system. Due to the navigation of sailors in the southern hemisphere, star catalogs now included the southern skies.

Four great star atlases made in the early 1600's and then 1 in 1690 revealed new information of the heavens. In 1603 Johann Bayer published Uranometria. Using Tycho Brahe's star catalog Bayer constructed large charts. His constellation figures, detailed and artistic filled one page each. The stars were accurately placed and the brightness of certain stars noted clearly in the text.

Modern Constellations

In 1928, the International Astronomical Union established 88 official constellations with clearly defined boundaries. A constellation now represents a group of stars, an area of the sky, and any star within the region belongs to only one constellation. Alpharatz belongs to Andromeda.

What are these 88 official constellations? You could make a little catalogue. There are 14 people, 9 birds, a couple of insects, 19 land animals, and so on in the list. There is a serpent, a dragon, a flying horse, there is even a river, and there are 29 inanimate objects – things like the Scales, the Plough and so forth.

Ancient astronomers and observers looked to the heavens and imagined great mythical creatures filling their night time skies. Navigators used these images to sail across vast oceans. Today we use these images and some of their ancient names to uniformly identify sections of the solar system that appear to us at night.

PLANETARY CARTOGRAPHY

Planetary Cartography forms an integral part of cartography and has found its manifestation in governmental activities, community efforts, professional organizations, and, in recent years, private activities. That activity substantiated with the revival of planetary exploration in the early 2000s when Europe visited the Moon for the first time, and the US launched a number of exploration mission. With the success of Asian spacecraft missions to the Moon and Mars joining the global planetary exploration endeavour, Planetary Cartography is increasingly becoming a global collaborative effort with Planetary Mapping being one of its main tools to accomplish the goals.

Planetary Cartography does not only provide the means to create science products after successful termination of a planetary mission by distilling data into maps. It also provides the basis to support planning (e.g., landing-site selection, orbit observation, traverse planning) and to facilitate mission conduct during the lifetime of a mission (e.g., observation tracking and hazard avoidance). After a mission's lifetime, information is stored in data archives – and eventually into maps and higher-level data products – to form a basis for research and for new scientific and engineering studies. The complexity of such tasks increases with every new dataset that has been put on this stack of in formation, and in the - same way as the complexity of autonomous probes increases, also tools that support these challenges require new levels of sophistication.

As of today, hundreds of planetary maps have been produced and published during a number of different frame-work programs and projects. Therein, different mapping efforts exist, either on a national level or as collaboration between groups participating as investigators in mapping missions. However, coordination of such tasks does not end with the compilation and publication of a set of maps. Coordination may be considered successfully only when mapping products have been provided to upcoming generations of researchers and mappers to allow efficient re-use of a new sustainable data basis. In order to accomplish this, mapping infrastructure, workflows, communication paths and validation tools have to be developed and made available.

Distillation of Information, Abstraction and Visualization

Visualization of data in general, and visualization of research data in particular, represents a simplified view on the real world, covering complex situations as well as the relationship between these. The process to accomplish this can be divided into four major parts: (1) data pre-processing and transformation, (2) visual mapping, (3) generation of views, and (4) perception/cognition. The mapping process in Planetary Cartography is comparable to established processes commonly employed in terrestrial cartographic workflows that are described as so-called visualization pipeline.

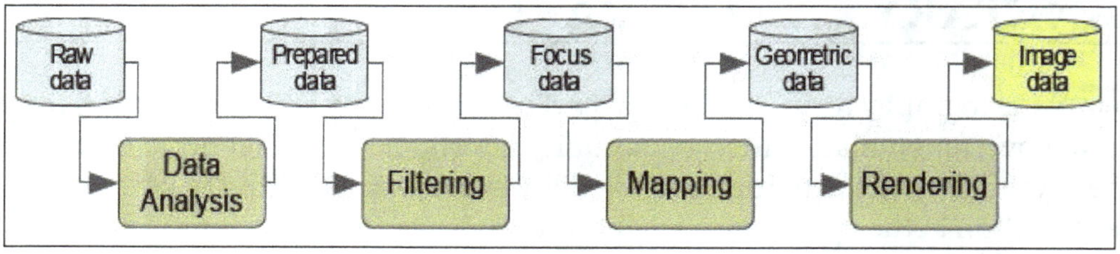

Visualization Pipeline.

This process is independent of actual production techniques and methods and describes different steps from acquisition and filtering of raw data (input), abstraction and generalization of information (distillation), and rendering, i.e. visualization of results (output). The last step in this pipeline is the recursion and by that the manipulation of previous processing steps: input and distillation.

Mapping of physical surfaces represents a special way of spatial data visualization within its scientific context as it does not show an objective visualization of a situation, but rather a subjective distinction of individual surface structures, units, and objects. This interpretation will generally be conducted by experts in their topical field whose expertise allows isolating, differentiating, and describing individual planetary objects based on remote-sensing data to reconstruct processes which shaped the current character of a surface. In planetary research and mapping, geographic information system technology (GIST) has been commonly employed to accomplish this since the early 2000s.

Challenges in Planetary Cartography

The standardization of cartographic methods and data products is critical for accurate and precise analysis and scientific reporting. This is more relevant today than ever before, as researchers have easy access to digital data as well as to the tools to process and analyze these various products. The life cycle of cartographic products can be short and standardized descriptions are needed to keep track of different development branches. One of our aims herein was to compartmentalize the processes of Planetary Cartography and to define, describe, and present the overall mapping process through its components breakdown.

Processes related to INPUT compartment cover all aspects related to the distillation of information that allows not only to produce higher-level products but also to create a basis for their stable representation and re-usability. One of the major tasks is to establish an international map-data basis by digitizing analogue maps and by establishing a uniform structure to describe existing data allowing them to be queried and accessed. For digital map products, a metadata description along with validation tools and platforms capable of providing access to archiving, distribution, and querying needs to be established. Standards already partially exist on a national level and some of the older higher-level map-data products are currently transferred to fit into such schemata. However, many map products exist all around the world and are distributed across different institutes. One task will be to review such products and to establish a methodological repertoire to transfer maps, to establish Mapping of physical surfaces represents a special way of common metadata scheme and to provide a common semantical basis.

Distillation processes ensure effective and efficient GIS-based management of data and derived map-data products in order to enable a consistent style of cartographic visualization. We identify three major tasks that are necessary to accomplish this: (1) definition and setup of rules and recommendations for GIS-based mapping process; (2) advocating for GIS-based implementation and distribution of international cartographic symbol standards; (3) generating generic, modular data models for GIS-based mapping and tools for transferring existing data into these models. Efforts are currently focusing on creating a templates-based framework on evaluation and optimization of existing map templates. In particular the short lifetime of products during on-going missions pose a considerable challenge when creating and operationally working with such models. Furthermore, recent work focuses on revising recommendations for cartographic symbols for geologic mapping. It encompasses critical review and updating of existing symbol standards for planetary geologic symbols.

Output processes cover all aspects of publishing and archiving mapping results in easily accessible archives and intuitive online interfaces and platforms. One method to achieve this is to incorporate already published maps along with their metadata into a uniform platform using a uniform format. That can be achieved by establishing metadata descriptions of maps and digital map projects that build on existing definitions and which can benefit from existing validation tools. The Planetary Data System in its version 4 has provided a flexible toolset to accomplish parts of this task in cooperation with the USGS ACS.

A next step is to build an internationally accessible digital map archive that includes digitized analogue maps and digital maps and mapping products. Already existing archives are provided by the USGS ACS or by the Planetary Data Set. The last point in this task group covers aspects of interoperability and exchange of map projects between different mapping and database systems. As different research institutes and individuals use different tools for mapping and data storage, procedures have to be found to allow conversions and also collaborative mapping in the future.

A general aim for this Planetary Cartography group is to work on concepts and approaches to foster future cooperation between cartographers and noncartographers. Also, collaboration should focus on reducing duplication of efforts and combining limited resources in order to address technical and scientific objectives. Recently - and primarily motivated by such tasks – international cross-collaborations between our institues, critical involvement within organizations like MAPSIT or ICA, but also on-going contribution to initiatives like VESPA have been established. This network will focus on (1) identifiying and prioritizing needs of the planetary catorgraphy community along with a a strategic timeline to accomplish prioritized goals, (2) keeping track of ongoing work across the globe in the field of Planetary Cartography, and (3) identifying areas of evolving technologies and innovation deal with mapping strategies as well as output media for the dissemination and communication of cartographic results.

CRITICAL CARTOGRAPHY

A critical cartography is the idea that maps – like other texts such as the written word, images or film – are not (and cannot be) value-free or neutral. Maps reflect and perpetuate relations of power, more often than not in the interests of dominant groups.

It is fairly easy to think of some ways in which maps embody power relations. One need not dig too deep within the history of mapping to see that they are intricately tied up in the history of nineteenth century colonialism and imperialism. Cartographers drew – and continue to draw – boundaries that separate people and resources. As another example, it is a fairly well-known fact that the commonly used Mercator projection of the globe is an inaccurate representation, because when cartographers 'flatten out' the spherical earth, they need to make certain choices: Size, shape and distances cannot all be maintained in the process. In the Mercator projection, the global North is vastly expanded at the expense of the South and Europe is placed squarely in the centre. As a further example, we may find it relatively easier – using an Ordinance Survey or Google Maps – to find a recently built supermarket than a longstanding squat, autonomous space, social centre or other radical space, or perhaps the site of the Battle of Hastings rather than the site of a historical radical struggle or riot. This does not just have practical implications for finding a space. Maps structure and limit our knowledge of the landscape, affecting our perception of what is important, the relative sizes and relationships between objects and spaces and where it is possible or safe to travel.

Critical Cartography is therefore, in the first instance, interested with theoretical critique of the social relevance, politics and ethics of mapping. The assumption that this is even a possibility – that maps are not simply neutral tools but rather strategic weapons that express power – leads to a second, practical, aspect of critical cartography. Groups and individuals at a grassroots level can also use mapping for a variety of purposes. Maps can be used to make counter-claims, to express competing interests, to make visible otherwise marginal experiences and hidden histories, to make practical plans for social change or to imagine utopian worlds.

It is important to note that maps are expressions not only of power, but of desire. Maps themselves can be objects of desire – some people enjoy looking at maps, or collecting historical ones. Maps also project our desires onto the landscape, they can map our hopes for the future, what we desire to see and that which we wish to ignore or hide. The process of mapping can also bring new ways of being and relating into the world, for example, we might experiment with new ways of organising and making decisions, such as non-hierarchy and consensus.

Academic literature tends to be fairly light on sketching alternative practices. There is a relatively large literature about 'counter-mapping', a practice which involves organisations such as NGOs and charities enabling indigenous communities to chart their territory in order to make land claims or protect resources from the encroachments of capital. These practices are undoubtedly progressive, but they have also been subject to criticism. They can involve representing communities' sometimes multiple and conflicting desires as a single representation, ignoring power differentials and exclusions within communities. This can be a necessary strategic act when attempting to make rights or resource claims to hierarchical entities such as states or trade organisations, yet can also help to perpetuate and legitimise such structures.

This is not to say that alternative mapping practices do not have a place in anarchist and non-hierarchical movements and studies. Social movements already use cartographies as ways of producing and communicating knowledge, yet these have rarely been studied in academic work. Examples of groups using counter-cartography include Bureau D'Etudes who produce huge geopolitical maps with massive amounts of information, highlighting for example links between corporations, financial institutions and arms trade companies, on a global scale. Other examples include the 56a info shop in London, which hosted a 'festival of mapping' in 2005. The info shop still has a huge archive of radical maps which can be visited by members of the public and members of the info shop still host radical mapping workshops and activities throughout London, the UK and worldwide.

Why is mapping a useful process for radical activists? Critical cartography can be a process of knowledge production and transformation. It is not just the 'final product' maps that are important; the process itself can involve learning together and producing new knowledge by bringing together multiple perspectives, by connecting different personal maps, or by creating collective maps through rotation, negotiation or consensus. Collaborative map-making can be a way to democratise knowledge-production. Mapping can also emphasise relations to institutions, landscapes, wildlife and environments, leading people to reconceive their relation to invisible structures or the natural world. More fundamentally it involves a reconfiguration of relations to space, dis-alienating one's relationship to space through the application of imagination.

The approach to mapping workshops has tended to resonate with the old Free Party slogan: "bring what you expect to find." Usually the approach is informal: We begin by getting people to think critically about mapping and alternative possibilities for mapping, using some of the ideas and examples listed above. We then try to facilitate the group in thinking through what kind of maps would be useful for their particular groups, and how they think it would be best to go about the process of mapping. We try to problematise some common dynamics that emerge – for example that people often veer towards wishing to map individually, yet one would hope that in a radical movement there would be some merit to mapping collectively.

Participants also often tend to parody traditional 'street map' styles and conventions, so it is sometimes worth thinking through the ways in which other environmental features which are often missed off conventional maps might be shown, or thinking about mapping non-visible aspects of the environment such as relationships, emotions or pollutants. It is also worth noting that maps need not be drawn on paper, nor need they be two-dimensional. Indigenous practices show possibilities for mapping such as textile pattern weaving, orally narrated storytelling and mythological maps, or maps that communicate using notches in sticks. The existence of multimodal and braille maps for blind people also point to some of the exclusionary aspects of visual mapping and possible alternatives.

The possibilities for mapping and map-making are as multiple as the people who choose to make maps.

The Process

- Who will be mapping, why and for what purpose?

- Does the group have common interests, values or desires?
- Is there a pre-decided theme, or will it be worked out as part of the process?
- Who and what will be invited and included, or perhaps implicitly excluded, and on what grounds?
- Is the space physically accessible to everyone who might attend, and can child-care be included if necessary?
- Are there any formal or informal hierarchies in the space, and how might these be addressed?
- Does the process itself produce any emotions or affects? Is it psychologically transformative?
- Who is the intended audience of the map?

The Map

- What will be mapped and why is this important?
- What materials or technology will be used?
- What will be made visible, or hidden, and why?
- What will be drawn, in what style, what colours?
- Are there any practical considerations for the map's intended use; e.g. should it be waterproof or capable of duplication?

The Life of the Map

- How will the map continue its life outside this space?
- How might the map function as a tool? Does it have any practical use?
- Who will be able to access, or might be excluded from using it, and how will it be used?
- What kind of knowledge is produced?
- Might the map trigger other cycles of learning/critique/mapping elsewhere?
- What are the political/ethical/social implications of these decisions?
- What changes or desires might the map bring into the world?

TERRAIN CARTOGRAPHY

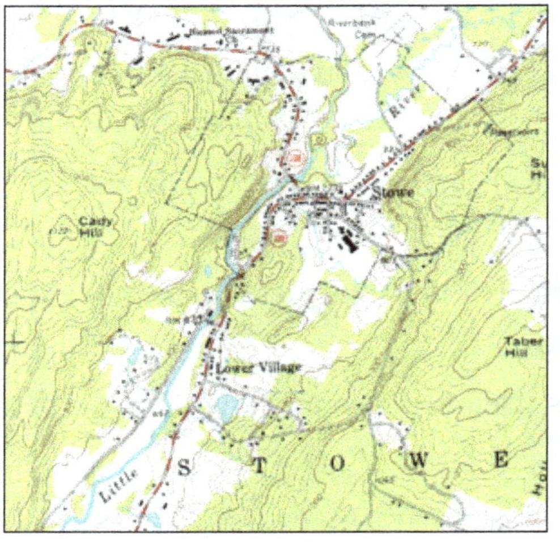

Topographic map of Stowe, Vermont with contour lines at 20-foot intervals.

Terrain or relief is an essential aspect of physical geography, and as such its portrayal presents a central problem in cartography, and more recently GIS and geovisualization.

The most obvious way to depict relief is to imitate it at scale, as in molded or sculpted solid terrain models and molded-plastic raised-relief maps. Because of the disparity between the horizontal and vertical scales of maps, raised relief is typically exaggerated.

On flat paper maps and computer screens, terrain can be depicted in a variety of ways.

Hill Profiles

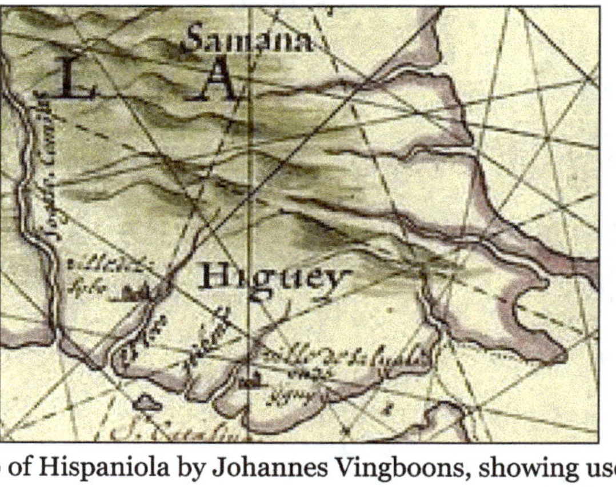

From a 1639 map of Hispaniola by Johannes Vingboons, showing use of hill profiles.

The most ancient form of relief depiction in cartography, hill profiles are simply illustrations of mountains and hills in profile, placed as appropriate on generally small-scale (broad area of coverage) maps. They are seldom used today except as part of an "antique" styling.

Hachures

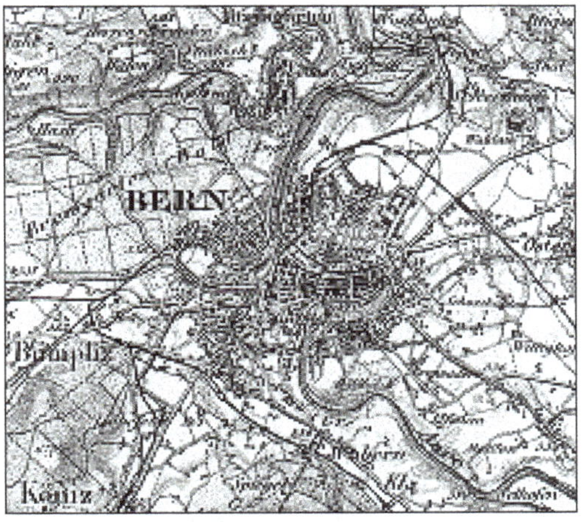

Dufour map of Bern; this is a shaded hachure map.

Hachures are also an older mode of representing relief. They are a form of shading, although different from the one used in shaded maps. They show the orientation of slope, and by their thickness and overall density they provide a general sense of steepness. Being non-numeric, they are less useful to a scientific survey than contours, but can successfully communicate quite specific shapes of terrain.

Hachure representation of relief was standardized by the Austrian topographer Johann Georg Lehmann in 1799.

Contour Lines

Contour lines (or isohypses) are isolines showing equal elevation. This is the most common way of numerically showing elevation, and is familiar from topographic maps.

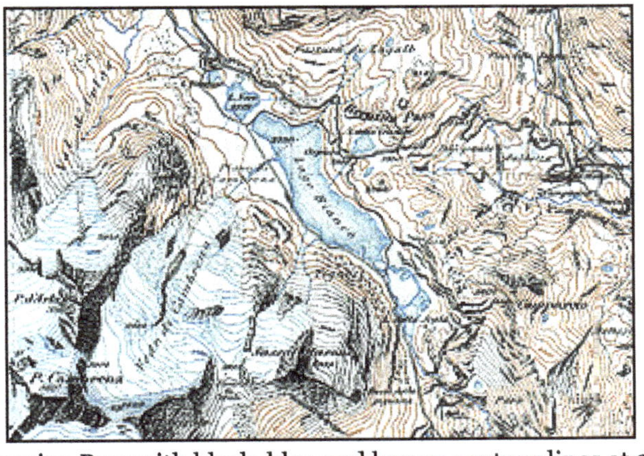

Siegfried map of Bernina Pass with black, blue and brown contour lines at 30-meter intervals.

Most 18th and early 19th century national surveys did not record relief across the entire area of coverage, calculating only spot elevations at survey points. The United States

Geological Survey (USGS) topographical survey maps included contour representation of relief, and so maps that show relief, especially with exact representation of elevation, came to be called topographic maps (or "topo" maps) in the United States, and the usage has spread internationally.

On maps produced by Swisstopo, the color of the contour lines is used to indicate the type of ground: black for bare rock and scree, blue for ice and underwater contours, brown for earth-covered ground.

Tanaka (Relief) Contours

The Tanaka (Relief) Contours technique is a method used to illuminate contour lines in order to help visualize terrain. Lines are highlighted or shaded depending on their relationship to a light source in the Northwest. If the object being illustrated would shadow a section of contour line, that contour would be represented with a black band. Otherwise, slopes facing the light source would be represented by white bands.

This method was developed by Professor Tanaka Kitiro in 1950, but had been experimented with as early as 1870, with little success due to technological limitations in printing. The resulting terrain at this point was a grayscale image. Cartographer Berthold Horn later created software to digitally produce Tanaka Contours, and Patrick Kennelly, another cartographer, later found a way to add color to these maps, making them more realistic.

There are a number of issues with this method. Historically, maps using Tanaka Contours could only be produced on gray backgrounds, which technology was unprepared for. This method is also very time consuming. In addition, the method takes into account mainly aspect and elevation, whereas slope is un accounted for resulting in a very terraced appearance.

Hypsometric Tints

Hypsometric tints (also called layer tinting, elevation tinting, elevation coloring, or hysometric coloring) are colors placed between contour lines to indicate elevation. These tints are shown as bands of color in a graduated scheme or as a color ramp applied to contour lines themselves. Hypsometric tinting of maps and globes is often accompanied by a similar method of bathymetric tinting to convey differences in water depth.

Shaded Relief

Shaded relief, or hill-shading, shows the shape of the terrain in a realistic fashion by showing how the three-dimensional surface would be illuminated from a point light source. The shadows normally follow the convention of top-left lighting in which the light source is placed near the upper-left corner of the map. If the map is oriented with north at the top, the result is that the light appears to come from the north-west. Although this is unrealistic lighting in the northern hemisphere, using a southern light source can cause multistable perception illusions, in which the topography appears inverted.

Top: map of Lake Mead area
Bottom: the same map with sun shading

Shaded relief was traditionally drawn with charcoal, airbrush and other artist's media. The Swiss master Eduard Imhof is widely regarded as the master of manual hill-shading technique and theory. Imhof's contributions included a multi-color approach to shading, with purples in valleys and yellows on peaks, which is known as "illuminated shading." Illuminating the sides of the terrain facing the light source with yellow colors provides greater realism, enhances the sense of the three-dimensional nature of the terrain, and make the map more aesthetically pleasing and artistic-looking.

Shaded relief is today almost exclusively computer-generated using digital elevation models (DEM), with a resulting different look and feel. Much work has been done in digitally recreating the work of Eduard Imhof, which has been fairly successful in some cases. The DEM may be converted to shaded relief using software such as Photoshop, QGIS, GRASS GIS or ArcMap's Spatial Analyst extension. Computer-generated effects to increase the realism of shaded relief include resolution bumping, the blending of hill-shades at different scales, and sky model shading, which simulates ambient light on terrain from many directions.

Physiographic Illustration

Pioneered by Erwin Raisz, this technique uses generalized texture to imitate landform shapes over a large area. A combination of hill profile and shaded relief, this style of terrain representation is simultaneously idiosyncratic to its creator—often hand-painted—and found insightful in illustrating geomorphological patterns.

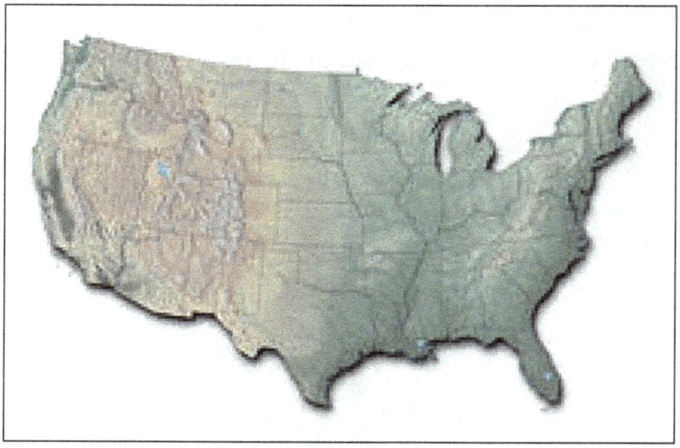

Web version of Patteron's Physical Map of the Coterminous United States.

More recently, Tom Patterson created a computer-generated map of the United States using Erwin Raisz's work as a starting point, the Physical Map of the Coterminous United States; he called the technique plan oblique relief.

ADVANTAGES AND DISADVANTAGES OF CARTOGRAPHY

Advantages of Cartography

- Less artistic skills are required to reproduce: Maps are a representation of a given area and are mostly dependent on symbols to represent the objects within the area. This means that no special artistic skills are required to reproduce the maps.
- Colors and patterns are easier to apply: Since there are very little artistic skills required to reproduce maps, it becomes easier to apply patterns and colors when representing features on a map.
- It is easier to make changes: Maps mainly rely on symbols and only show specific features of a given area as opposed to everything in that area. This means it is easier to make changes in case anything changes or when a feature was not properly represented.
- Integration is easy: Maps can easily be integrated with other applications and used concurrently. This also makes it easier to read and understand maps especially when reading in relation to other applications.
- Easy conversion of maps projections: Cartography makes it easy to make conversions of map projections.

- Maps can be maintained and updated much faster: Cartography has made it easier to maintain maps and update them much faster especially when it comes to digital maps.
- Cartography gives true representation of the region in a short scale: Cartography makes it easy to give a representation of a specific area on a very small piece of paper thanks to the scaling abilities of cartography.
- Easier data symbolization: Cartography has made it is easy to symbolize data on a piece of paper to make sense in a real world environment. This makes it easier to represent any data or information within limited space and time.
- Work with styles and symbols: Cartography allows users to work with styles and symbols to represent objects on a piece of paper.
- Graphical data representation is possible: It is possible to represent data both symbolically and graphically thanks to cartographic capabilities.

Disadvantages of Cartography

- Cartography is prone to human error: Since maps are designed and drawn by humans, it is possible to paste human error hence giving distorted information.
- Shapes may be distorted: Maps, especially the printed maps on paper may give distorted information or the shapes and objects may easily be compromised.
- The scale is variable hence it is not a true map: Maps depend on variation in scale to represent a large area. This is not a true representation of the real world area and may give incomplete information.
- No map is ever completely accurate or complete: Although cartography has enabled the representation of areas on maps, there is no guarantee that maps are complete or accurate. There will always be some level of error or incompleteness in the representations.
- No map can show all the features found in an area at any given time: There is no map that shows 100 percent of the features found in a given area. All maps lack certain degree of information or data.
- Maps only show relevant features and veils the unnecessary features: Cartography has allowed maps to show only specific features that are important to users at a specific time and not all the features that are available in the area.
- Maps require interpretation: Maps mostly use symbols to represent objects on land and these require interpretation. This means map reading requires some skills that may not be available to all.

- Maps are susceptible to bias: Maps are drawn and designed by different entities who may have certain biases towards certain areas hence not give full information regarding the area.
- Maps mostly use symbols to represent objects: Printed maps mostly rely on symbols to represent data. These symbols are sometimes not clear.
- Maps may be difficult to understand for the visually impaired: Map reading requires vision which may be difficult for the visually impaired.

RELEVANCE OF CARTOGRAPHY

In the geospatial domains, we can witness that more spatial data than ever is produced currently. Numerous sensors of all kinds are available, measuring values; storing them in databases, which are linked to other databases being embedded in whole spatial data infrastructures; following standards and accepted rules. We can witness also that we are not short of ever more new modern technologies for all parts of the spatial data handling processes, including data acquisition (e.g., unmanned aerial vehicles currently), data modeling (e.g., service-oriented architectures, cloud computing), and data visualization and dissemination (e.g., location-based services, augmented reality). So where are we now with all those brave, new developments?

Obviously, we are not short of data in many ways. Clearly, we can state that it is rather the opposite. The problem is often not that we don't have enough data but rather too much. We need to make more and more efforts to deal with all that data in an efficient sense, mining the relevant information and linking and selecting the appropriate information for a particular scenario. This phenomenon is being described as "big data." Often, application developments start there. Because we have access to data, we make something with it. We link it, we analyze it, we produce applications out of it. We call this a data-driven approach.

We are also not short of technologies. It is rather the opposite; while just being able to fully employ the potential of a particular data acquisition, modeling, or dissemination technology, new technologies come in and need to be considered. New technologies become available more and more quickly and need to be evaluated, addressed, and applied. Often, application development starts there. Because we have a new technology available, we make something with it. We call this a technology-driven approach.

However, the particular need, demand, question, or problem of a human user is often taken into account only when the data-driven or technology-driven application, product, or system has been built. Often, this causes problems or leads to products, systems, and applications that are not accepted, not efficient, or simply not usable. By starting from the question. What are the demands, questions, problems, or needs of human users in respect to location? we could eventually apply data and technology in a sense that they serve such user-centered approaches rather than determine the use.

But how can we better unleash the big potential of geoinformation in such truly interdisciplinary approaches? How can we make sure that spatial data is really applicable for governments, for decision makers, for planners, for citizens through applications, products, and systems, which are not forcing them to adapt to the system but are easy to use and efficiently support the human user?

Maps can be seen as the perfect interface between a human user and big data.

In this respect, maps and cartography play a key role. Maps are most efficient in enabling human users to understand complex situations. Maps can be understood as tools to order information by their spatial context. Maps can be seen as the perfect interface between a human user and all that big data and thus enable human users to answer location-related questions, to support spatial behavior, to enable spatial problem solving, or simply to be able to become aware of space.

Today, maps can be created and used by any individual stocked with just modest computing skills from virtually any location on earth and for almost any purpose. In this new mapmaking paradigm, users are often present at the location of interest and produce maps that address needs that arise instantaneously. Cartographic data may be digitally and wirelessly delivered in finalized form to the device in the hands of the user or the requested visualization derived from downloaded data in situ. Rapid advances in technologies have enabled this revolution in mapmaking by the millions. One such prominent advance includes the possibility to derive maps very quickly immediately after the data has been acquired by accessing and disseminating maps through the Internet. Real-time data handling and visualization are other significant developments, as well as location-based services, mobile cartography, and augmented reality.

While the above advances have enabled significant progress on the design and implementation of new ways of map production over the past decade, many cartographic principles remain unchanged, the most important one being that maps are an abstraction of reality. Visualization of selected information means that some features present in reality are depicted more prominently than others, while many features might not even be depicted at all. Abstracting reality makes a map powerful, as it helps to understand and interpret very complex situations very efficiently.

Abstraction is essential. Disaster management can be used as an example to illustrate the importance and power of abstract cartographic depictions. In the recovery phase, quick production of imagery of the affected area is required using depictions that allow the emergency teams to understand the situation on the ground from a glance at the maps. Important on-going developments supporting the rescue work in the recovery phase are map derivation technologies, crowdsourcing and neocartography techniques and location-based services. The role of cartography in the protection phase of the disaster management cycle has always been crucial. In this phase, risk maps are produced, which enable governors, decision makers, experts, and the general public alike to understand the kind and levels of risk present in the near and distant surroundings. Modern cartography enables the general public to participate in the modeling and visualizing of the risks neighbourhoods may suffer from on a voluntary basis. Modern cartography also helps to quickly disseminate crucial information.

In this sense, cartography is most relevant. Without maps, we would be "spatially blind." Knowledge about spatial relations and location of objects are most important to learn about space, to act in space, to be aware of what is where and what is around us, or simply to be able to make good decisions. Cartography is also most contemporary, as new and innovative technologies have an important impact into what cartographers are doing. Maps can be derived automatically from geodata acquisition methods, such as laser scanning, remote sensing, or sensor networks. Smart models of geodata can be built allowing in-depth analysis of structures and patterns. A whole range of presentation forms are available nowadays, from maps on mobile phones all the way to geoinformation presented as augmented reality presentations.

The successful development of modern cartography requires integrated, interdisciplinary approaches from such domains as computer science, communication science, human-computer interaction, telecommunication sciences, cognitive sciences, law, economics, geospatial information management, and cartography. It is those interdisciplinary approaches that make sure that we work toward human-centered application developments by applying innovative engineering methods and tools in a highly volatile technological framework. A number of important technology-driven trends have a major impact on what and how we create, access, and use maps, creating previously unimaginable amounts of location-referenced information and thus putting cartographic services in the center of the focus of research and development.

Where are we heading? What we can expect in the near future is that information is available anytime and anywhere. In its provision and delivery, it is tailored to the user's context and needs. In this, the context is a key selector for which and how information is provided. Cartographic services will thus be widespread and of daily use in a truly ubiquitous manner. Persons would feel spatially blind without using their map-based services, which enable them to see who or what is near them, get supported and do searches based on the current location, and collect data on-site accurately and timely. Modern cartography applications are already demonstrating their huge potential and change how we work, how we live, and how we interact.

References

- Cartography, entry: newworldencyclopedia.org, Retrieved 8 January, 2019
- Node, geog160: psu.edu, Retrieved 13 May, 2019
- Design-principles-for-cartography, mapping, product, products, arcgis-blog: esri.com, Retrieved 6 January, 2019
- Celestial-cartography, chapter, atd-fscj-introastronomy: lumenlearning.com, Retrieved 5 August, 2019
- Celestial-cartography-maps-of-the-constellations, astronomy: osr.org, Retrieved 30 March, 2019
- Advantages-and-disadvantages-of-cartography, geography: grindgis.com, Retrieved 1 April, 2019
- Planetary-Cartography-Activities-and-Current-Challenges: researchgate.net, Retrieved 2 July, 2019
- Jenny, Bernhard; Hurni, Lorenz (2006). “Swiss-Style Colour Relief Shading Modulated by Elevation and by Exposure to Illumination”. The Cartographic Journal. 43 (3): 198–207. Doi:10.1179/000870406X158164

2

Elements and Concepts of Cartography

Some of the major elements and concepts related to cartography are symbology, map legend, contour line, scale bars, compass rose, map projection, cartographic labeling, chorography, cartographic generalization, photogrammetry, etc. The topics elaborated in this chapter will help in gaining a better perspective about these elements and concepts.

GEOGRAPHIC COORDINATE SYSTEMS

A geographic coordinate system (GCS) uses a three-dimensional spherical surface to define locations on the earth. A GCS is often incorrectly called a datum, but a datum is only one part of a GCS. A GCS includes an angular unit of measure, a prime meridian, and a datum (based on a spheroid).

A point is referenced by its longitude and latitude values. Longitude and latitude are angles measured from the earth's center to a point on the earth's surface. The angles often are measured in degrees (or in grads). The following illustration shows the world as a globe with longitude and latitude values.

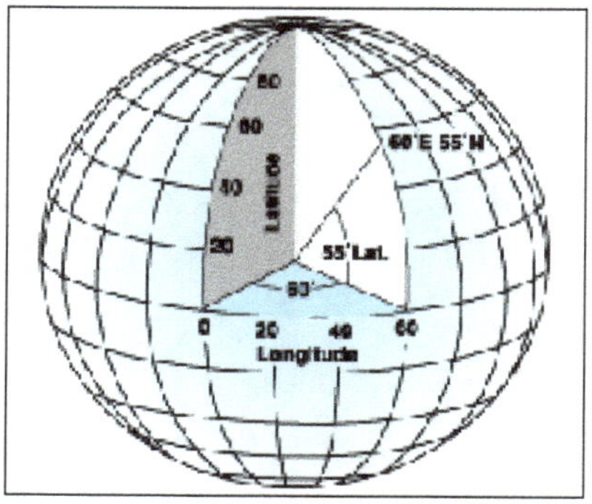

In the spherical system, horizontal lines, or east–west lines, are lines of equal latitude, or parallels. Vertical lines, or north–south lines, are lines of equal longitude, or meridians. These lines encompass the globe and form a gridded network called a graticule.

The line of latitude midway between the poles is called the equator. It defines the line of zero latitude. The line of zero longitude is called the prime meridian. For most geographic coordinate systems, the prime meridian is the longitude that passes through Greenwich, England. Other countries use longitude lines that pass through Bern, Bogota, and Paris as prime meridians. The origin of the graticule (0,0) is defined by where the equator and prime meridian intersect. The globe is then divided into four geographical quadrants that are based on compass bearings from the origin. North and south are above and below the equator, and west and east are to the left and right of the prime meridian.

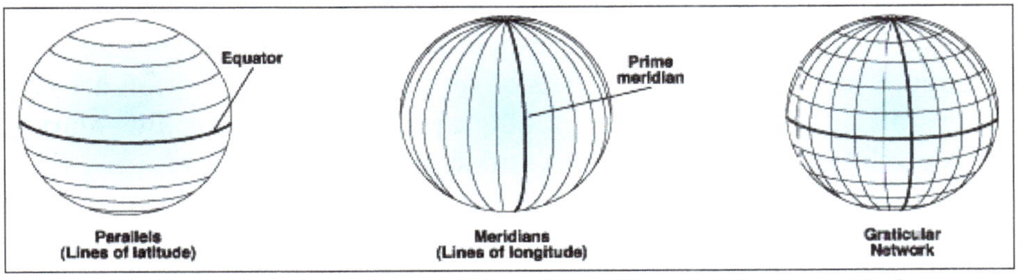

This illustration shows the parallels and meridians that form a graticule.

Latitude and longitude values are traditionally measured either in decimal degrees or in degrees, minutes, and seconds (DMS). Latitude values are measured relative to the equator and range from -90° at the South Pole to +90° at the North Pole. Longitude values are measured relative to the prime meridian. They range from -180° when traveling west to 180° when traveling east. If the prime meridian is at Greenwich, then Australia, which is south of the equator and east of Greenwich, has positive longitude values and negative latitude values.

It may be helpful to equate longitude values with X and latitude values with Y. Data defined on a geographic coordinate system is displayed as if a degree is a linear unit of measure. This method is basically the same as the Plate Carrée projection.

Although longitude and latitude can locate exact positions on the surface of the globe, they are not uniform units of measure. Only along the equator does the distance represented by one degree of longitude approximate the distance represented by one degree of latitude. This is because the equator is the only parallel as large as a meridian. (Circles with the same radius as the spherical earth are called great circles. The equator and all meridians are great circles.)

Above and below the equator, the circles defining the parallels of latitude get gradually smaller until they become a single point at the North and South Poles where the meridians converge. As the meridians converge toward the poles, the distance represented by one degree of longitude decreases to zero. On the Clarke 1866 spheroid, one degree of longitude at the equator equals 111.321 km, while at 60° latitude it is only 55.802 km. Because degrees of latitude and longitude don't have a standard length, you can't measure distances or areas accurately or display the data easily on a flat map or computer screen.

CARTOGRAPHIC CONSIDERATIONS

Size

The size at which a map is going to be printed or viewed dictates a number of things, the most important of which is scale. Once the scale has been decided, the level of generalisation can also be determined. In cartography 'generalisation' means "how much detail can be shown on the map face before it gets too busy", or more simply "how cluttered do I want the map to be?"

In this modern world, maps are either printed onto paper or are viewed electronically (e.g. on computer screen). The principals of cartographic composition are essentially the same for both in that they can be photographically/electronically enlarged or reduced but the relationship between text and symbols remains the same. For printed maps, if more, or less, detail is required then a new map needs to be produced or enlargement(s) added to the map. For web/electronic maps it is possible to have a range of views with different scales and detail as one zooms in or out.

Position on the Paper

A map is positioned within the framework of its 'piece of paper'. It is an important part of the map maker's art to decide how the map will be positioned on the piece of paper. There are no set rules and very few conventions in regard to this process – often referred to as Map Layout.

The map may:

- Occupy the entire 'piece of paper' with the marginalia being included on the face of the map.
- Occupy only a part of the 'piece of paper' with the marginalia occupying the remainder.
- The map face and the marginalia may be completely separate from each other – either as a separate document or printed on the back of the map. This occurs where the marginalia is very complex. However, be warned, where the marginalia is separate to the map there is considerable risk that the map would become unreadable if the two are separated.

Where a map is to be bound into a publication such as an Atlas, the map may be placed off–centre on the page (allowing for paper which may be lost when binding occurs) – this is called a binding edge.

These three maps are good examples of these map layout concepts. The first map is a General Reference map of Australia, where the map occupies the entire 'piece of paper' and the marginalia is included on the face of the map. Note how easy to read this is and how the colours 'draw your eye' to the map itself.

The second map is a more complex map, of soil types in northern Tasmania. Here the marginalia has been placed in a separate panel to the actual map. This option has been used as the map itself is very complex and the legend is also very complex, because it has to explain in detail the information that is being shown. In this example the map maker could easily have decided to print the legend separately or on the back of the map – because it is very complex.

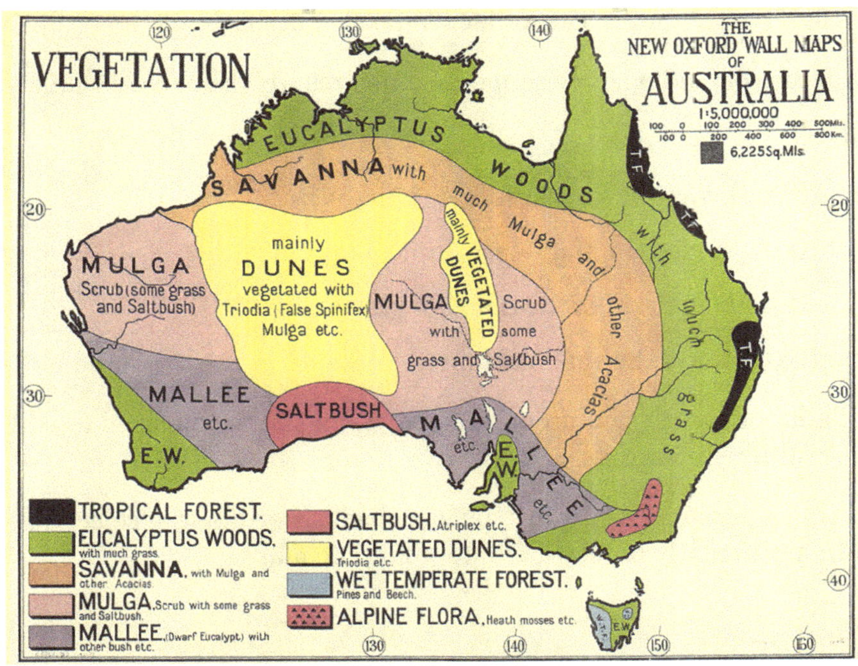

The third map of Australia is a simple map of vegetation types. Because the map is clear and of a well-known part of the world the location information is limited, but it has all the important elements of a map – especially a Title, Legend, Scale and Production Documentation. This map is also placed to one side of the 'piece of paper' – this is to allow for the loss of paper when it is bound into a book. However, it could equally have been positioned like this because the map maker liked the 'look of it'.

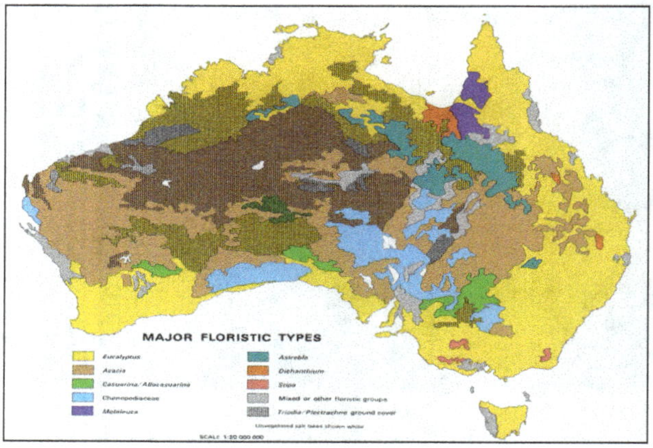

Text

In general, all text (both on the face of the map and in the marginalia) should use the same font. The smallest font should be easily read (the simple 'rule of thumb' is that the smallest character should be no less that 0.8mm high). Occasionally for special effect or emphasis an elaborate font is used for the Title or for symbols – but this is not normally the case.

To assist with ease of reading, the number of font sizes which are used should be limited.

For example:

- Map Title (always the largest font) – e.g. 16pt.
- Any text on the face of map – e.g. 10pt to 14pt.
- Any text in the margin of the map – e.g. 10pt.
- Footnotes (always the smallest font) – e.g. 8pt.

Italics, bold and differently coloured fonts may (and should) be used to add emphasis or distinguish different features.

Over the years elaborate sets of rules have evolved regarding the selection and placement of text – particularly in relation to topographic maps.

Symbols

If possible, it is best to use symbols which are well known – national standards and conventions should be adopted where applicable. These can also be adapted to suit the map's needs. If special symbols are developed for a map they should be included in the map legend.

Developing a symbol usually means defining its colour, line weights and shape. This can be very complex or quite simple depending on the requirements of the map. In general, the simpler the symbol the more intuitive it becomes and the easier the map it is to read.

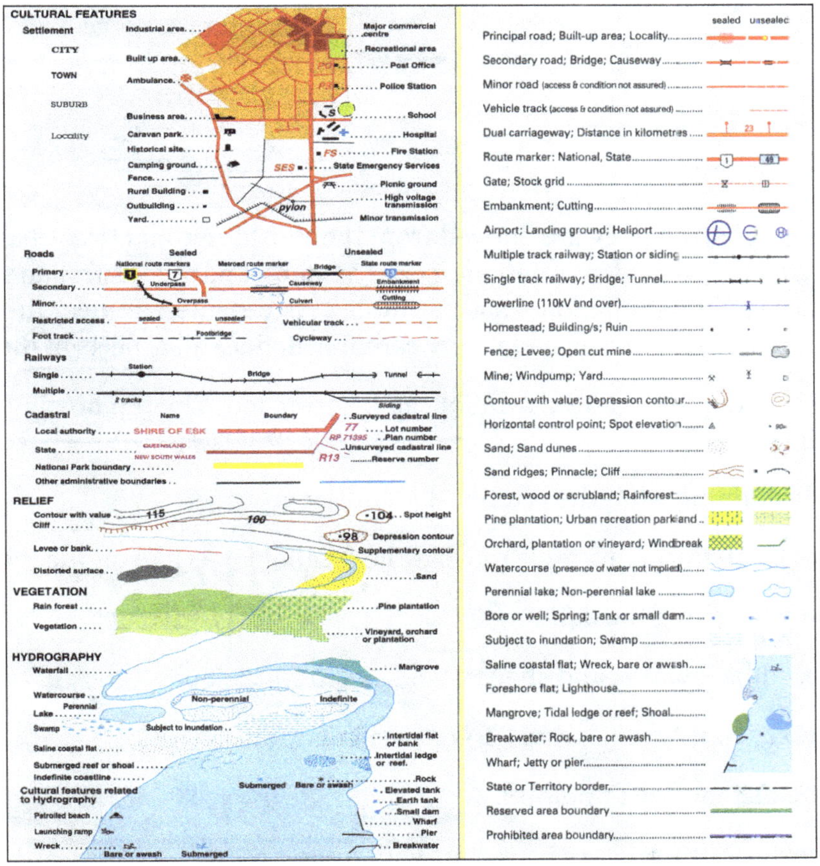

These two legends are from topographic maps which have different scale, different series and different producers. They show how similar, but not identical symbols have been used for the same feature. Rivers are blue; roads are red and vegetation is green. But note some of the differences. For example; contours are brown on one and black on the other; while different types of railway lines are symbolised differently.

Getting a Little Technical about Symbols

Many map makers, especially those who are doing a 'one–off' map, don't worry too much when designing their symbols. They simply choose a colour and style out of the 'paint box' and use it.

However in series maps it is important that each individual map sheet looks the same. For these, rules need be developed about how a symbol is drawn and used – especially if a map is to be produced using computers. This is set of definitions is commonly called the maps symbology and they are usually included in the specifications for the map series.

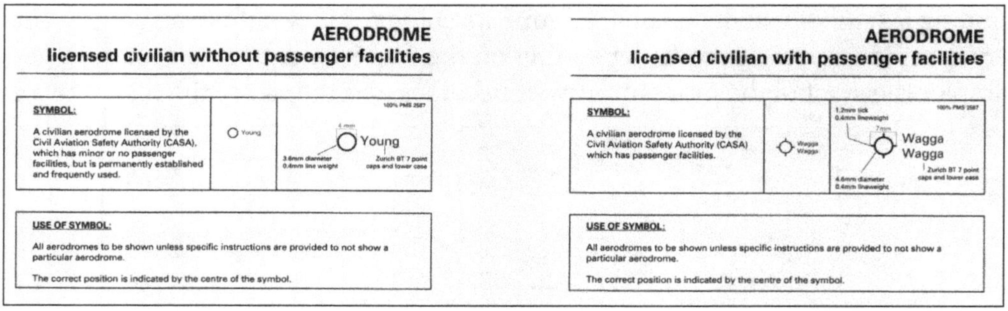

These examples of symbology are derived from the World Aeronautical Chart Specifications for Australia. This is an international series of maps, and the Australian specification is a refinement of the international specification. It is designed to meet the needs map production systems in Australia and it very carefully defines all aspects of the symbols.

These are two examples of how a symbol can be designed. The top box in both images is defining the symbol by:

- Firstly defining what the symbol is identifying.
- Secondly giving an example of what the symbol looks like on the map.
- Thirdly by giving the defining.
- Line widths, lengths and shapes.
- The text's font type (Zurich BT is a particular font).
- The text's font size (7 points).
- The text's font format (words start with a capital).
- Colours that are to be used, such as a brand name of a particular colour.

Colours and Patterns

Colour design is a science in itself but a few basic rules can help a map achieve its aims and still be nice to look at:

- A consistent set of colours should be used for symbols (e.g. roads in red, rivers in blue).
- Use tones/shades and different colours to differentiate and emphasise.
- The colour selected, its intensity, and the size of the area will have a large bearing on the visual impact of the map. For example bright 'solid' (100% colour) yellow might be a good colour for a small area but a light yellow (say 15% of the solid colour) would often be a better option for a very large area.

- Use colour and tonal variation to give 'light and shade' to the map, ie use the full spectrum of colours, and make some 'dark' and others 'light' so that the map doesn't look flat.
- Avoid using patterns, especially 'hachures' (a series of parallel lines) as they can copy or print badly and may obscure important underlying detail. (Patterns are often necessary and suitable for thematic maps such as geological maps.)

If your map is likely to be copied in a single colour (e.g. a black and white photocopy) then simple patterns should be used instead of subtle colour differences. As with colours, the pattern chosen as well as its density will determine the aesthetic impact of the map. Heavier patterns (such as closely spaced lines) will generally have more visual effect than an open dot for example. Variation, and above all moderation, is the key to successful patterning for a monochrome map.

Symbols and Symbology

This is a distinction with can be very confusing to the uninitiated and the difference is very simple:

- A symbol is a shape which is defined — e.g. the shape of an anchor.
- Symbology is the set of rules which relate to how that symbol is used — e.g.:
    - An anchor which is up–side down means a ship wreck.
    - A blue anchor means it is a safe place to moor a boat.
    - A red anchor means it is not a safe place to moor a boat etc.

Ownership and Logos

AUSTRALIA

© Commonwealth of Australia 2000
No part of this publication may be reproduced, stored in a retrieval system or transmitted in any form or by any means, electronic, mechanical, photocopying, recording or otherwise.

In most cases map makers wish to retain the ownership of their map (called copyright). This ownership can relate to the information shown on the map (often called the intellectual property) as well as the map itself. This is done by a statement which may include the copyright symbol (©) and possibly the displaying of an organisations logo or other official 'trademarks'.

As well as showing ownership, logos can also be used record sponsorship or responsibility for the information being mapped. It is important that map makers remember the following:

- Logos cannot be used without formal approval.
- There may be in–house rules as to how and where logos and trademarks are displayed.
- Avoid using excessive numbers of logos – these can overpower the map.

While most maps have ownership statements attached to them, they may not have use restrictions. However, if a new map is made using an existing map as all or part of its source it is common for to require that the original map be identified.

Maps produced by Australian Government agencies, the ownership is stated by:

- Displaying the Australian Coat of Arms.
- Adding the words "© Commonwealth of Australia".

It may also list the uses and restrictions on the map.

Elevation

Many maps show elevation – especially topographic maps.

In the Topographic Maps section, elevation is usually shown using contour lines. In simplistic terms, a contour line is a line which joins points of equal elevation. Where these lines are above sea level they are simply called contour lines and where they are below sea level they are called bathymetric contour lines.

One convention is that contour lines are generally shown as brown or black lines and bathymetric contour lines are shown as blue lines.

Another convention is that numbers are placed on a selection of the contour line to indicate their height above or below sea level. These are positioned with the top of the number pointing up the slope – so by looking at only one number you know which way is 'uphill' and which way is 'downhill'.

In the 1800's with the adoption of contours as the preferred method to show elevation, map makers experimented with ways to further enhance the maps. Two methods have become the modern standard:

- Relief shading
- Hypsometric tints

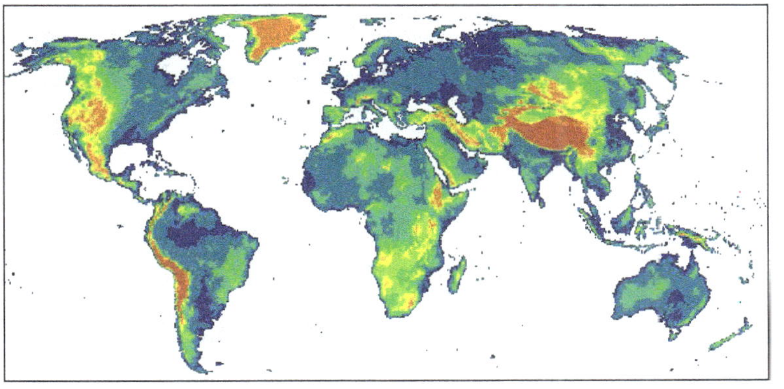

Bathymetry and Hypsometry

Hypsometry is the measurement of the height of the land.

Bathymetry is the measurement of the depth of underwater features.

In simple terms one is about mapping the land above sea level and the other charting the land below sea level.

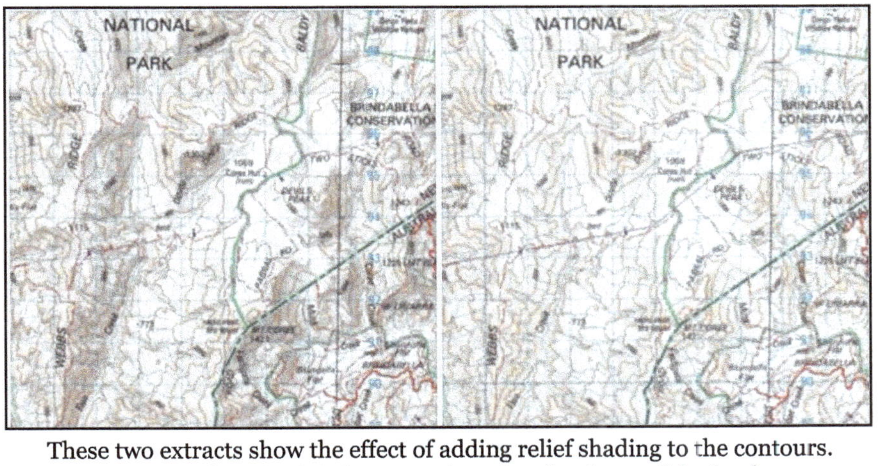

These two extracts show the effect of adding relief shading to the contours.
It is much easier it is for you to imagine the shape of the land.

Relief Shading (or Hill Shading)

This technique was created to give a flat map a 3D visual impression. It is designed to simulate the effect of the shadow cast by relief features such as mountains, ranges and gorges. It is based on the convention that an imaginary light source (eg the Sun) is in the northwest corner of the map and the shadows spread out towards the southeast corner.

Originally this work was undertaken by highly skilled map makers, but today automated computer packages are able to achieve the same effect.

Hypsometric and Bathymetric Tints

This technique was also created to give a flat map a 3D visual impression. However it is much easier to produce than the relief shading technique and has become very popular for General Reference maps – especially for Atlases.

Over time a convention has emerged:

- Soft colours are used for elevations/depths near sea level.
- Stronger colours are used for elevations/depths further away from sea level.
- For the land these colours are 'earth hues' (browns) and for the sea these colours are 'water hues' (blues).
- In some parts of the world, where elevations are significantly higher than Australia's, greys and white (perhaps to suggest snow) are often used for very high land (e.g. for peaks in the Himalayas).

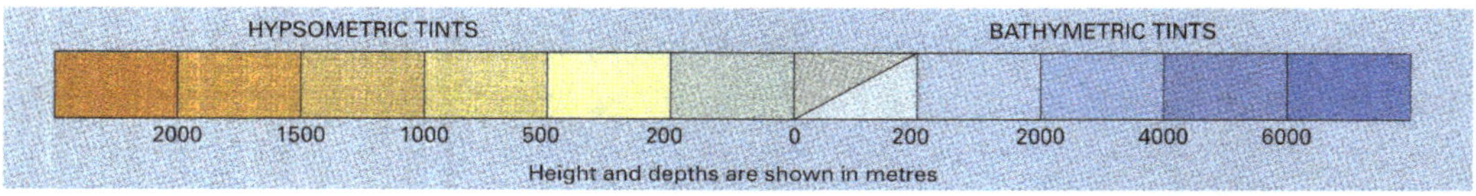

Firstly, please note that for the scale bar for hypsometric tints the contour interval is not even – the contour intervals being 200, 300 then 500 metres for the land and 200, 1800 then 2000 for the sea. These differences are to allow for the significant differences in elevation/depth between the two.

Secondly, note that for the 0–200 metres below sea level two colours are given – the top one is for height below sea level on the land (eg in Lake Eyre) and the bottom one is for depth below sea level in the sea.

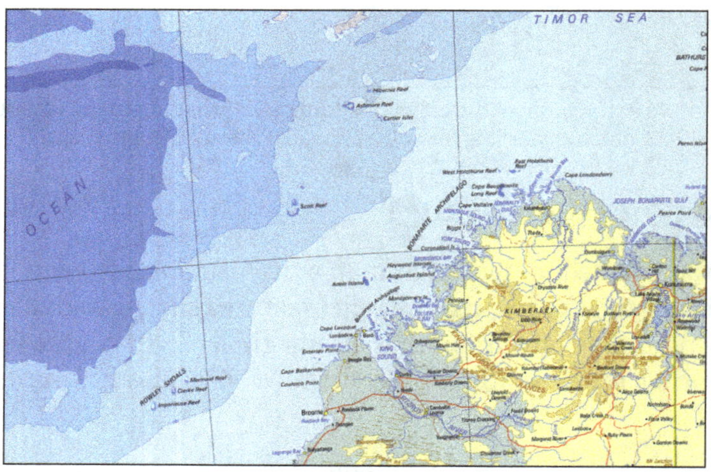

This is part of a General Reference Map of Australia. The use of colour instantly implies height and depth and gives the map a 3D look-and-feel.

Enlargement and Inset

An inset is a small map which is included with the main map, but it is from a different area. These may be at the same or different scales to the main map.

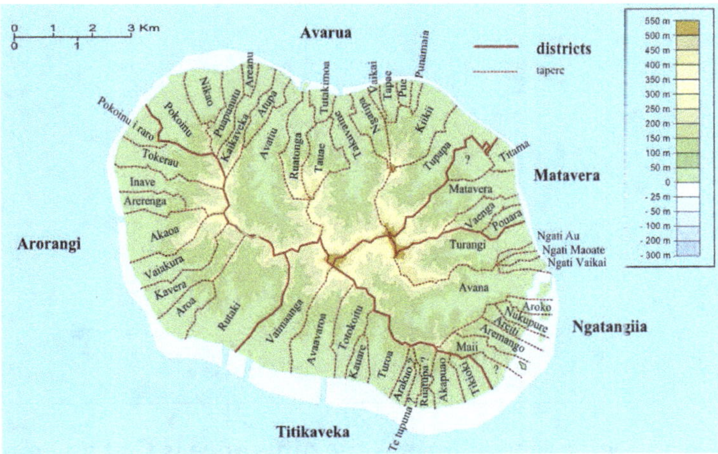

Rarotonga Island in the Cook Islands.

In the case of the map of Rarotonga – the maps in the north–west and south–west corners are inset and they are at a smaller scale than the main map. They are designed to show where Rarotonga is located relative to other islands and countries in its vicinity.

In the case of the Maritime Zones report map (below) the inset is at the same scale, but it is from an area outside the extent of the main map.

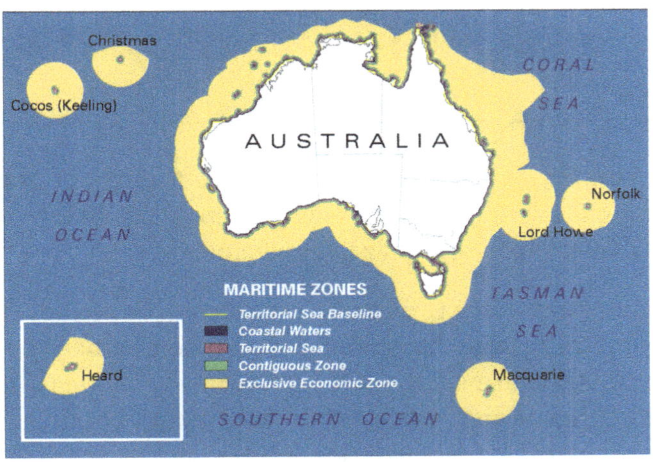

It is not uncommon for a map to be composed entirely of different scales insets and not have a main map. An excellent example of this is maps of scattered islands (as is found in the Pacific Ocean) – these are typically (but not exclusively) found it world atlases.

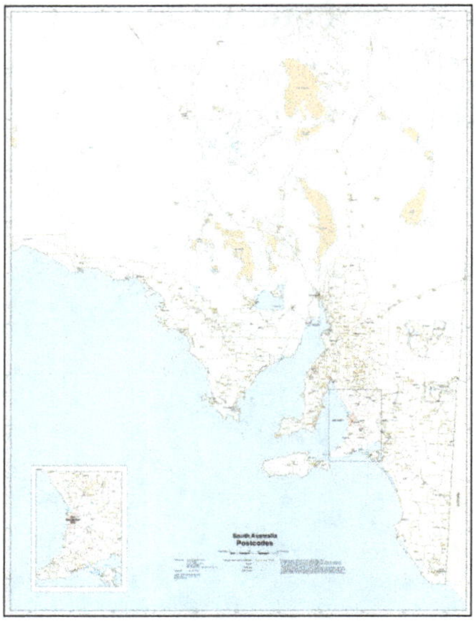

An enlargement is a specific type of inset – the difference is that an enlargement shows a part of the main map that is already being shown. These areas are usually more congested than the main map and are therefore shown a second time at a larger scale than the main map. This is done to supply greater detail of areas that have a great deal more to show than the bulk of the main map.

For example in the map of Rarotonga, the entire island is shown at one scale and the city of Avarua (which is also shown in the main body of the map) is shown in more detail by using an enlargement.

SYMBOLOGY

Symbology, in the context of Cartographic design, is the use of graphical techniques to represent geographic information on a map. Map symbols for geographic features include Visual variablessuch as color, size, and shape.

Generically, a symbol is an object, picture, written word, sound, or particular mark that represents something else by association, resemblance, or convention. For example, Roman numerals are symbols for quantitative values and personal names are symbols representing individual people. On a map, a red cross is a commonly understood symbol to indicate the location of a hospital, crossed sabres may indicate the site of a battlefield, and a blue region would commonly be interpreted as a water body. Semiotics is the scientific and philosophical study of how symbols work by establishing these connections between the representation and the represented concepts and real-world features.

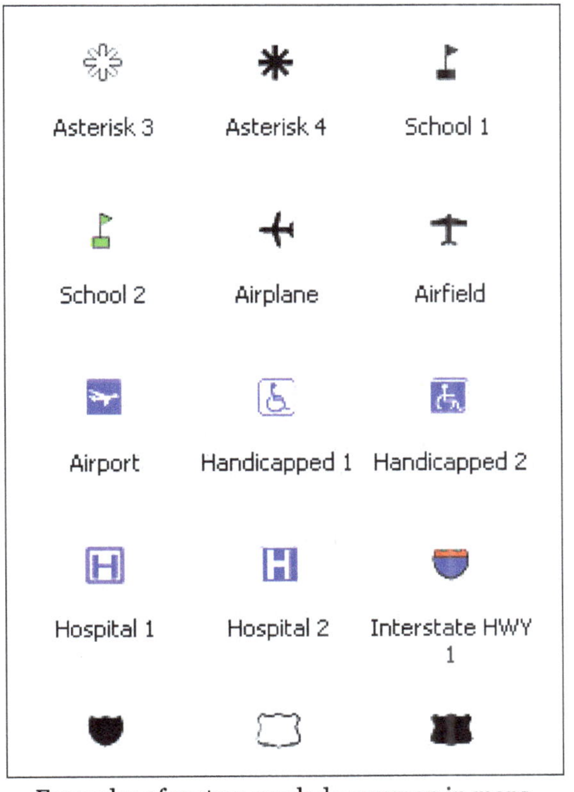

Examples of custom symbols common in maps.

Map Symbols

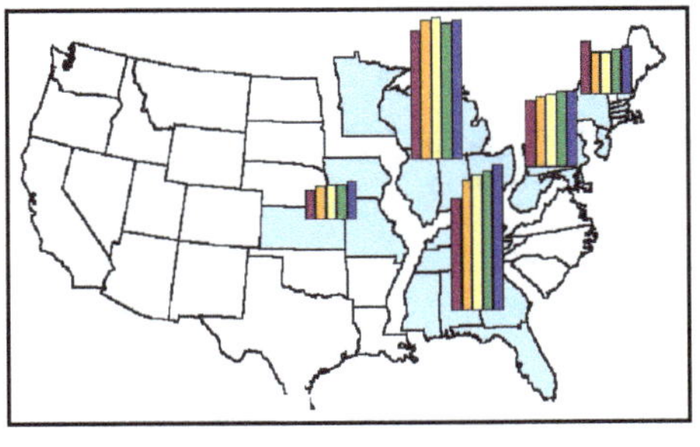

A chart used as a point symbol to portray quantitative data.

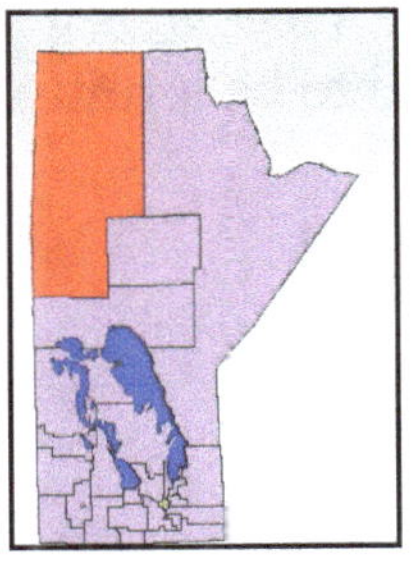

Color used a region symbol to portray a quantitative attribute.

Maps communicate their messages through symbols-drawn graphics that represent spatial phenomena such as objects, places, or attributes. At their most basic, map graphics can be categorized by Dimension: points, lines, and regions; each can be portrayed using symbology. These symbols are commonly used to describe different features mapped. For example, cities or airports are commonly represented as point symbols (depending on scale), roads or railroads are usually represented by line symbols and the cities, lakes, or forests are common examples of region symbols.

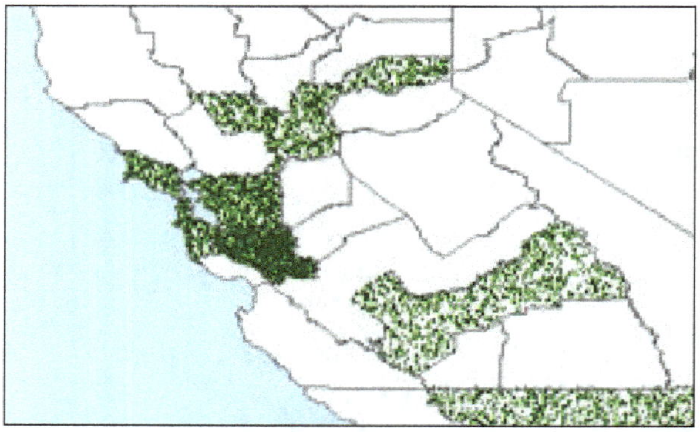

Texture (dot density) used as a region symbol to portray a quantitative attribute.

Map symbols are created by controlling Visual variables such as color, shape, and size; the range of such variables was set forth by Jaques Bertin and subsequent cartographers. When designing a map, the cartographer determines that a certain combination of these variables-a symbol (e.g., a dashed 0.5pt blue line)-represents a certain class of geographic feature (e.g., an intermittent stream). Although there is no set standard on symbology for all maps, especially among thematic maps, various conventions (such as using contour lines for elevation or blue for water) are commonly followed for some classes of maps.

The choice of symbology is a crucial part of cartographic design; the goal is for the map reader to spend less time figuring out what the symbols mean, and thus more time using the symbols to understand the World. A good symbol is easily recognizable (i.e., connected to the geographic features and concepts it represents), is aesthetically pleasing, and works in harmony with other symbols (for example, as part of a clear visual hierarchy). A legend is needed to explain the meaning of the symbols that cannot safely be assumed to be intuitive.

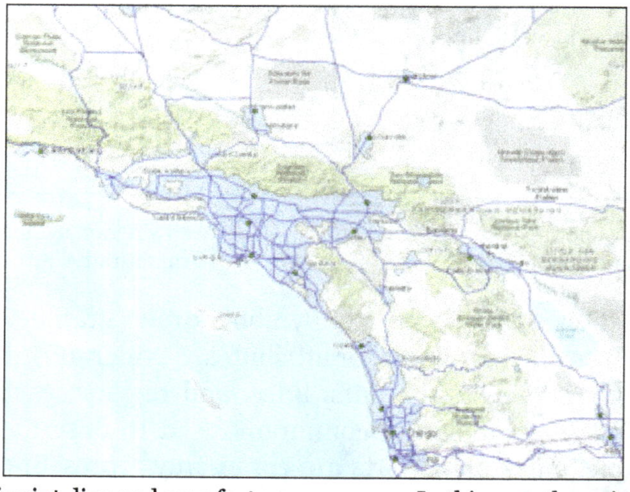

Examples of point, line and area features on a map. In this map the point represents populated places, the line represents roads, and the area represents urban areas.

Types of Map Symbols

According to semiotics, map symbols are "read" by map users when they make a connection between the graphic mark on the map (the *sign*), a general concept (the *interpretant*), and a particular feature of the real world (the *referent*). For example, thick blue line (*sign*) = major river (*interpretant*), and *this* thick blue line = The Colorado River (*referent*). Map symbols can thus be categorized by how they suggest this connection:

- Image or Iconic Symbols *look like* the real-world feature, although it is often in a generalized manner. For example, a tree icon to represent a forest, or green denoting vegetation.
- Functional Symbols directly represent the *activity* that takes place at the represented feature. For example, a picture of a skier to represent a ski resort or a tent to represent a campground.
- Conceptual Symbols directly represent a *concept* related to the represented feature. For example, a dollar sign to represent an ATM, or a Star of David to represent a Jewish synagogue.
- Conventional Symbols do not have any intuitive relationship but are so *commonly used* that map readers eventually learn to recognize them. For example, a red line to represent a highway or a cross to represent a hospital.
- Ad Hoc Symbols do not have any intuitive or commonly accepted relationship; in fact, the symbol is often used only for a single map. For example, a polygon in a choropleth map filled with a particular shade of red that means "high income." These symbols require effective legendsand/or labels to be correctly interpreted.

Symbology Standards

Cartographers typically have a great degree of freedom to design and implement map symbology according to their personal tastes, creativity, and innovation. However, in some applications, the need for consistency and immediate usability has driven the development of standard sets of symbols.

Military Symbology

Military symbology has a very large and detailed library of symbols for use in military cartography, Intelligence, and Engineering. Each branch of America's armed forces use the same symbols making it easy to identify units that are friendly, neutral, or enemy. Also, military symbols are used to show equipment, installations, military or civilian, and military operations on and around the battle space. All these are a large part of military symbology. The army gets its symbology from NATO. NATO develops and approves the symbols that the armed services use. The Allied Procedural Publication (APP) are NATO's standards for military map-making symbology. An example of military symbols are APP-6A.

Emergency Mapping Symbology

Emergency Mapping Symbology are specialized sets of symbols used by various organizations when planning for or responding to emergencies. These emergencies can be naturally caused (tsunami, earthquake, tornado, etc.) or human caused (rioting, terrorism, hijacking, etc.). Currently there is no international standard for emergency mapping symbology which has meant that various nations have created their own national symbology set. Recognized and standardized symbol sets help create a Common Operating Picture (COP) for varied organizations that have been brought together during a crisis or emergency. Symbols that are easy to identify with and easy to distribute are seen as key elements in creating maps that can be used to reduce fatalities, injuries or loss of property.

MAP LEGEND

Maps and charts use stylized shapes and symbols as well as common map colors to designate features such as mountains, highways, and cities. The legend is a small box or table on the map that explains the meanings of those symbols. The legend may also include a map scale for help in determining distances.

Designing a Map Legend

If you are designing a map and legend, you may come up with your own symbols and colors or you may rely on standard sets of icons, depending on the purpose of your illustration. Legends usually appear near the bottom of a map or around the outer edges. They may be placed outside or within the map. If placing the legend within the map, set it apart with a distinctive frame or border and don't cover up any important portions of the map.

While the style can vary, a typical legend has a column with the symbol followed by a column describing what that symbol represents.

- Position the legend on the map and include all the symbols used on the map in the legend.
- Keep the legend (and the map) as simple and easy to read as possible. This isn't the time for a fancy design.
- The style of the legend should match the style of the map itself.

Creating the Map

Before you create the legend, you need the map. Maps are complex graphics. The designer's challenge is to make them as simple and clear as possible without omitting any important information. Most maps contain the same types of elements, but a designer controls how they are presented visually. Those elements include:

- Title,
- Legend,
- Scale,
- Mapped areas (objects, land, water),
- Borders,
- Symbols,
- Labels.

As you work in your graphics software, use layers to separate the different types of elements and to organize what can end up being a complicated file. Complete the map before you prepare the legend.

Symbol and Color Selection

You don't have to reinvent the wheel with your map and legend. It may be best for your reader if you don't. Highways and roads are usually represented by lines of various widths, depending on the size of the road, and are accompanied by interstate or route labels. Water is usually indicated by the color blue. Dashed lines indicate borders. An airplane indicates an airport.

Examine your symbols' fonts. We may already have what we need for our map, or we can search online for a map font or a PDF that illustrates the various map symbols.

Be consistent in the use of symbols and fonts throughout the map and legend — and simplify. The goal is to make the map and legend reader-friendly, useful, and accurate.

CONTOUR LINE

A contour line (also isoline, isopleth, or isarithm) of a function of two variables is a curve along which the function has a constant value, so that the curve joins points of equal value. It is a plane section of the three-dimensional graph of the function $f(x, y)$ parallel to the x, y plane. In cartography, a contour line (often just called a "contour") joins points of equal elevation (height) above a given level, such as mean sea level. A contour map is a map illustrated with contour lines, for example a topographic map, which thus shows valleys and hills, and the steepness or gentleness of slopes. The contour interval of a contour map is the difference in elevation between successive contour lines.

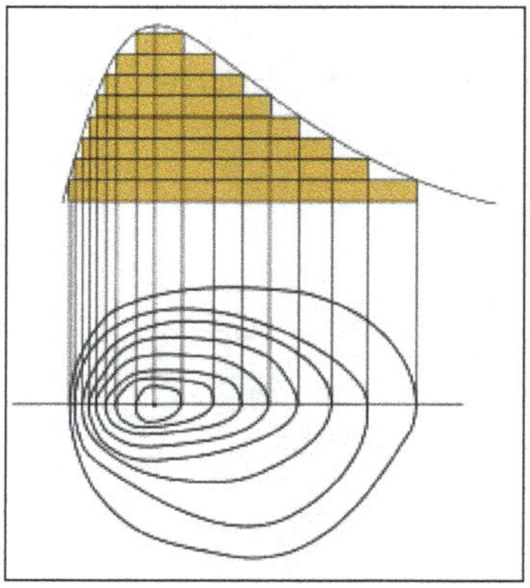

The bottom part of the diagram shows some contour lines with a straight line running through the location of the maximum value. The curve at the top represents the values along that straight line.

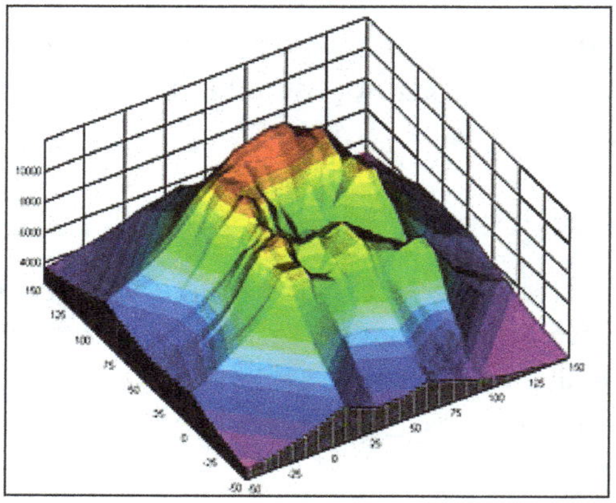

A three-dimensional surface, whose contour graph is below.

More generally, a contour line for a function of two variables is a curve connecting points where the function has the same particular value.

The gradient of the function is always perpendicular to the contour lines. When the lines are close together the magnitude of the gradient is large: the variation is steep. A level set is a generalization of a contour line for functions of any number of variables.

Contour lines are curved, straight or a mixture of both lines on a map describing the intersection of a real or hypothetical surface with one or more horizontal planes. The configuration of these contours allows map readers to infer relative gradient of a parameter and estimate that parameter at specific places. Contour lines may be either traced on a visible three-dimensional model of the surface, as when a photogrammetrist viewing a stereo-model plots elevation contours, or interpolated from estimated surface elevations, as when a computer program threads contours through a network of observation points of area centroids. In the latter case, the method of interpolation affects the reliability of individual isolines and their portrayal of slope, pits and peaks.

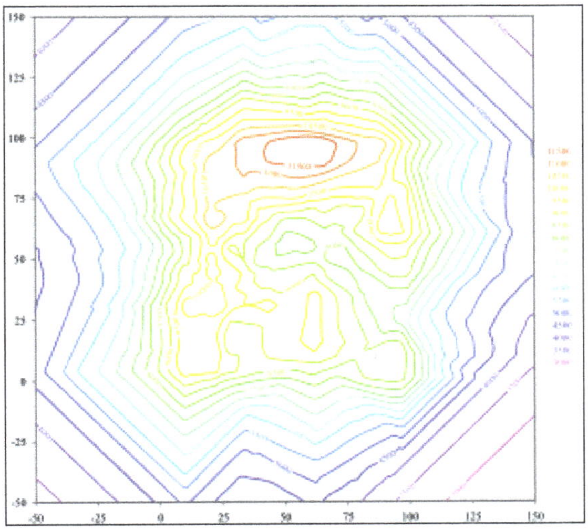

A two-dimensional contour graph of the three-dimensional surface in the above picture.

Types

Contour lines are often given specific names beginning "iso-" according to the nature of the variable being mapped, although in many usages the phrase "contour line" is most commonly used. Specific names are most common in meteorology, where multiple maps with different variables may be viewed simultaneously. The prefix "iso-" can be replaced with "isallo-" to specify a contour line connecting points where a variable changes at the same *rate* during a given time period.

The words *isoline* and *isarithm* are general terms covering all types of contour line. The word *isogram* was proposed by Francis Galton in 1889 as a convenient generic designation for lines indicating equality of some physical condition or quantity; but it commonly refers to a word without a repeated letter.

An isogon is a contour line for a variable which measures direction. In meteorology and in geomagnetics, the term *isogon* has specific meanings. An isocline is a line joining points with equal slope. In population dynamics and in geomagnetics, the terms *isocline* and *isoclinic line* have specific meanings.

Equidistant Points

A curve of equidistant points is a set of points all at the same distance from a given point, line, or polyline. In this case the function whose value is being held constant along a contour line is a distance function.

Isopleths

In geography, the word *isopleth* is used for contour lines that depict a variable which cannot be measured at a point, but which instead must be calculated from data collected over an area. An example is population density, which can be calculated by dividing the population of a census district by the surface area of that district. Each calculated value is presumed to be the value of the variable at the centre of the area, and isopleths can then be drawn by a process of interpolation. The idea of an isopleth map can be compared with that of a choropleth map.

In meteorology, the word *isopleth* is used for any type of contour line.

Meteorology

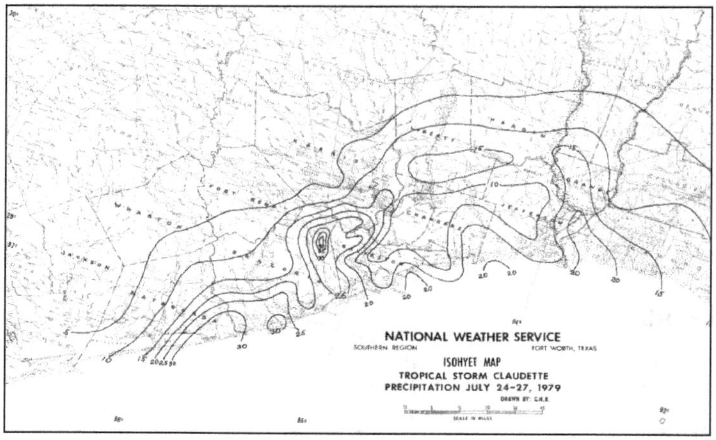

Isohyetal map of precipitation.

Meteorological contour lines are based on interpolation of the point data received from weather stations and weather satellites. Weather stations are seldom exactly positioned at a contour line (when they are, this indicates a measurement precisely equal to the value of the contour). Instead, lines are drawn to best approximate the locations of exact values, based on the scattered information points available.

Meteorological contour maps may present collected data such as actual air pressure at a given time, or generalized data such as average pressure over a period of time, or forecast data such as predicted air pressure at some point in the future.

Thermodynamic diagrams use multiple overlapping contour sets (including isobars and isotherms) to present a picture of the major thermodynamic factors in a weather system.

Barometric Pressure

An isobar is a line of equal or constant pressure on a graph, plot, or map; an isopleth or contour line of pressure. More accurately, isobars are lines drawn on a map joining places of equal average atmospheric pressure reduced to sea level for a specified period of time. In meteorology, the barometric pressures shown are reduced to sea level, not the surface pressures at the map locations. The distribution of isobars is closely related to the magnitude and direction of the wind field, and can be used to predict future weather patterns. Isobars are commonly used in television weather reporting.

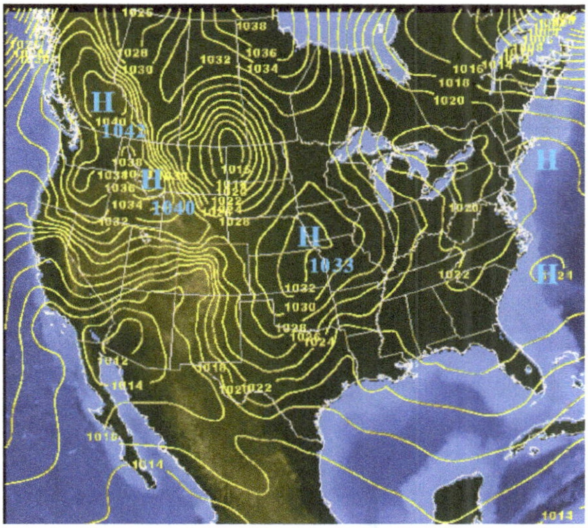

Isobars.

Isallobars are lines joining points of equal pressure change during a specific time interval. These can be divided into *anallobars*, lines joining points of equal pressure increase during a specific time interval, and *katallobars*, lines joining points of equal pressure decrease. In general, weather systems move along an axis joining high and low isallobaric centers.Isallobaric gradients are important components of the wind as they increase or decrease the geostrophic wind.

An isopycnal is a line of constant density. An *isoheight* or *isohypse* is a line of constant geopotential height on a constant pressure surface chart. Isohypse and isoheight are simply known as lines showing equal pressure on a map.

Temperature and Related Subjects

An isotherm is a line that connects points on a map that have the same temperature. Therefore, all points through which an isotherm passes have the same or equal temperatures at the time indicated. An isotherm at 0 °C is called the freezing level. The term was coined by the Prussian geographer and naturalist Alexander von Humboldt, who as part of his research into the geographical distribution of plants published the first map of isotherms in Paris, in 1817.

An isogeotherm is a line of equal mean annual temperature. An isocheim is a line of equal mean winter temperature, and an isothere is a line of equal mean summer temperature.

An isohel is a line of equal or constant solar radiation.

The 10 °C (50 °F) mean isotherm in July, marked by the red line, is commonly used to define the border of the Arctic region.

Rainfall and Air Moisture

An isohyet or isohyetal line is a line joining points of equal rainfall on a map in a given period. A map with isohyets is called an isohyetal map. An isohume is a line of constant relative humidity, while an isodrosotherm is a line of equal or constant dew point. An isoneph is a line indicating equal cloud cover.

An isochalaz is a line of constant frequency of hail storms, and an isobront is a line drawn through geographical points at which a given phase of thunderstorm activity occurred simultaneously. Snow cover is frequently shown as a contour-line map.

Wind

An isotach is a line joining points with constant wind speed. In meteorology, the term isogon refers to a line of constant wind direction.

Freeze and Thaw

An isopectic line denotes equal dates of ice formation each winter, and an isotac denotes equal dates of thawing.

Physical Geography and Oceanography

Elevation and Depth

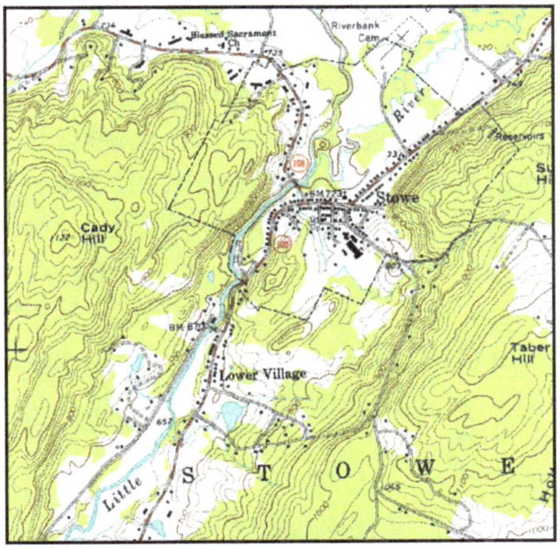

Topographic map of Stowe, Vermont. The brown contour lines represent the elevation. The contour interval is 20 feet.

Contours are one of several common methods used to denote elevation or altitude and depth on maps. From these contours, a sense of the general terrain can be determined. They are used at a variety of scales, from large-scale engineering drawings and architectural plans, through topographic maps and bathymetric charts, up to continental-scale maps.

"Contour line" is the most common usage in cartography, but isobath for underwater depths on bathymetric maps and isohypsefor elevations are also used.

In cartography, the contour interval is the elevation difference between adjacent contour lines. The contour interval should be the same over a single map. When calculated as a ratio against the map scale, a sense of the hilliness of the terrain can be derived.

Interpretation

There are several rules to note when interpreting terrain contour lines:

- The rule of Vs: Sharp-pointed vees usually are in stream valleys, with the drainage channel passing through the point of the vee, with the vee pointing upstream. This is a consequence of erosion.
- The rule of Os: Closed loops are normally uphill on the inside and downhill on the outside, and the innermost loop is the highest area. If a loop instead represents a depression, some maps note this by short lines called hachures which are perpendicular to the contour and point in the direction of the low. (The concept is similar to but distinct from hachures used in hachure maps.)
- Spacing of contours: Close contours indicate a steep slope; distant contours a

shallow slope. Two or more contour lines merging indicates a cliff. By counting the number of contours that cross a segment of a stream, the stream gradient can be approximated.

Of course, to determine differences in elevation between two points, the contour interval, or distance in altitude between two adjacent contour lines, must be known, and this is normally stated in the map key. Usually contour intervals are consistent throughout a map, but there are exceptions. Sometimes intermediate contours are present in flatter areas; these can be dashed or dotted lines at half the noted contour interval. When contours are used with hypsometric tints on a small-scale map that includes mountains and flatter low-lying areas, it is common to have smaller intervals at lower elevations so that detail is shown in all areas. Conversely, for an island which consists of a plateau surrounded by steep cliffs, it is possible to use smaller intervals as the height increases.

Electrostatics

An isopotential map is a measure of electrostatic potential in space, often depicted in two dimensions with the electostatic charges inducing that electric potential. The term equipotential line or isopotential line refers to a curve of constant electric potential. Whether crossing an equipotential line represents ascending or descending the potential is inferred from the labels on the charges. In three dimensions, equipotential surfaces may be depicted with a two dimensional cross-section, showing equipotential lines at the intersection of the surfaces and the cross-section.

The general mathematical term level set is often used to describe the full collection of points having a particular potential, especially in higher dimensional space.

Magnetism

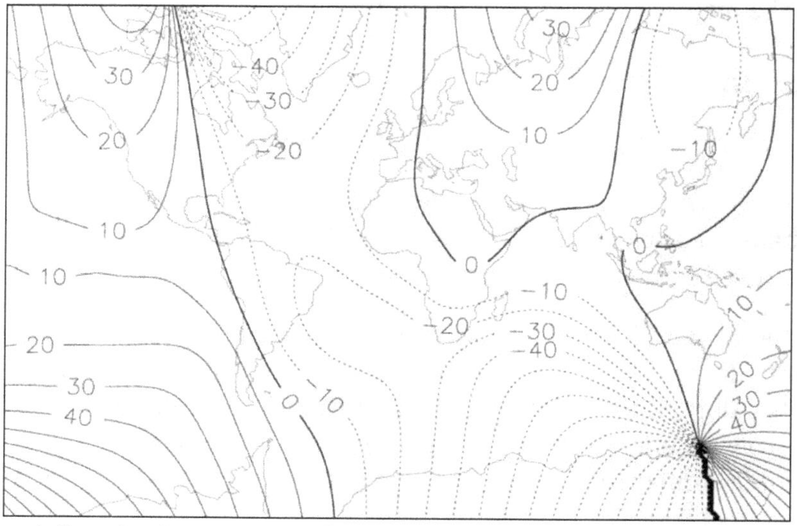

Isogonic lines for the year 2000. The agonic lines are thicker and labeled with "0".

In the study of the Earth's magnetic field, the term isogon or isogonic line refers to a line of constant magnetic declination, the variation of magnetic north from geographic north. An agonic line is drawn through points of zero magnetic declination. An isoporic line refers to a line of constant annual variation of magnetic declination.

An isoclinic line connects points of equal magnetic dip, and an aclinic line is the isoclinic line of magnetic dip zero. An isodynamic line connects points with the same intensity of magnetic force.

Oceanography

Besides ocean depth, oceanographers use contour to describe diffuse variable phenomena much as meteorologists do with atmospheric phenomena. In particular, isobathytherms are lines showing depths of water with equal temperature, isohalinesshow lines of equal ocean salinity, and Isopycnals are surfaces of equal water density.

Geology

Various geological data are rendered as contour maps in structural geology, sedimentology, stratigraphy and economic geology. Contour maps are used to show the below ground surface of geologic strata, fault surfaces (especially low angle thrust faults) and unconformities. Isopach maps use isopachs (lines of equal thickness) to illustrate variations in thickness of geologic units.

Environmental Science

In discussing pollution, density maps can be very useful in indicating sources and areas of greatest contamination. Contour maps are especially useful for diffuse forms or scales of pollution. Acid precipitation is indicated on maps with isoplats. Some of the most widespread applications of environmental science contour maps involve mapping of environmental noise (where lines of equal sound pressure level are denoted isobels), air pollution, soil contamination, thermal pollution and groundwater contamination. By contour planting and contour ploughing, the rate of water runoff and thus soil erosion can be substantially reduced; this is especially important in riparian zones.

Ecology

An isoflor is an isopleth contour connecting areas of comparable biological diversity. Usually, the variable is the number of species of a given genus or family that occurs in a region. Isoflor maps are thus used to show distribution patterns and trends such as centres of diversity.

Social Sciences

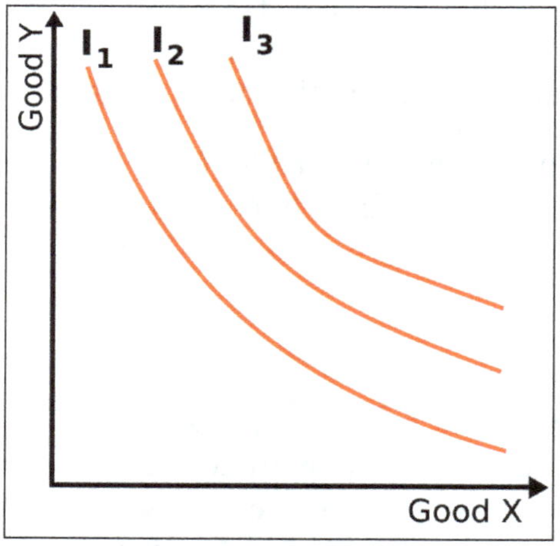

From economics, an indifference map with three indifference curves shown. All points on a particular indifference curve have the same value of the utility function, whose values implicitly come out of the page in the unshown third dimension.

In economics, contour lines can be used to describe features which vary quantitatively over space. An isochrone shows lines of equivalent drive time or travel time to a given location and is used in the generation of isochrone maps. An isotimshows equivalent transport costs from the source of a raw material, and an isodapane shows equivalent cost of travel time.

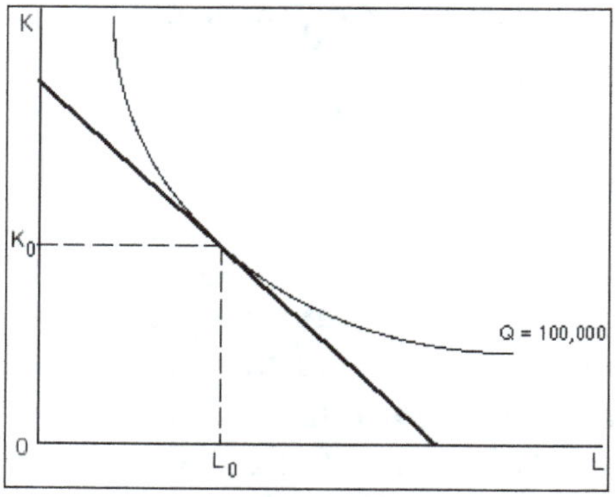

A single production isoquant (convex) and a single isocost curve (linear). Laborusage is plotted horizontally and physical capital usage is plotted vertically.

Contour lines are also used to display non-geographic information in economics. Indifference curves (as shown at left) are used to show bundles of goods to which a person would assign equal utility. An isoquant (in the image at right) is a curve of equal production quantity for alternative combinations of input usages, and an isocost curve (also in the image at right) shows alternative usages having equal production costs.

In political science an analogous method is used in understanding coalitions (for example the diagram in Laver and Shepsle's work).

In population dynamics, an isocline shows the set of population sizes at which the rate of change, or partial derivative, for one population in a pair of interacting populations is zero.

Statistics

In statistics, isodensity lines or isodensanes are lines that join points with the same value of a probability density. Isodensanes are used to display bivariate distributions. For example, for a bivariate elliptical distribution the isodensity lines are ellipses.

Thermodynamics, Engineering and Other Sciences

Various types of graphs in thermodynamics, engineering, and other sciences use isobars (constant pressure), isotherms (constant temperature), isochors (constant specific volume), or other types of isolines, even though these graphs are usually not related to maps. Such isolines are useful for representing more than two dimensions (or quantities) on two-dimensional graphs. Common examples in thermodynamics are some types of phase diagrams.

Isoclines are used to solve ordinary differential equations. In interpreting radar images, an isodop is a line of equal Doppler velocity, and an isoecho is a line of equal radar reflectivity.

Other Phenomena

- Isochasm: Aurora equal occurrence,
- Isochor: Volume,
- Isodose: Absorbed dose of radiation,
- Isophene: Biological events occurring with coincidence such as plants flowering,
- Isophote: Illuminance,
- Mobile telephony: Mobile received power and cell coverage area.

Algorithms

- Finding boundaries of level sets after image segmentation:
  - Edge detection,
  - Level-set method,
  - Boundary tracing.
- Active contour model.

Graphical Design

To maximize readability of contour maps, there are several design choices available to the map creator, principally line weight, line color, line type and method of numerical marking.

Line weight is simply the darkness or thickness of the line used. This choice is made based upon the least intrusive form of contours that enable the reader to decipher the background information in the map itself. If there is little or no content on the base map, the contour lines may be drawn with relatively heavy thickness. Also, for many forms of contours such as topographic maps, it is common to vary the line weight and color, so that a different line characteristic occurs for certain numerical values. For example, in the topographic map above, the even hundred foot elevations are shown in a different weight from the twenty foot intervals.

Line color is the choice of any number of pigments that suit the display. Sometimes a sheen or gloss is used as well as color to set the contour lines apart from the base map. Line colour can be varied to show other information.

Line type refers to whether the basic contour line is solid, dashed, dotted or broken in some other pattern to create the desired effect. Dotted or dashed lines are often used when the underlying base map conveys very important (or difficult to read) information. Broken line types are used when the location of the contour line is inferred.

Numerical marking is the manner of denoting the arithmetical values of contour lines. This can be done by placing numbers along some of the contour lines, typically using interpolation for intervening lines. Alternatively a map key can be produced associating the contours with their values.

If the contour lines are not numerically labeled and adjacent lines have the same style (with the same weight, color and type), then the direction of the gradient cannot be determined from the contour lines alone. However, if the contour lines cycle through three or more styles, then the direction of the gradient can be determined from the lines. The orientation of the numerical text labels is often used to indicate the direction of the slope.

Plan View versus Profile View

Most commonly contour lines are drawn in plan view, or as an observer in space would view the Earth's surface: ordinary map form. However, some parameters can often be displayed in profile view showing a vertical profile of the parameter mapped. Some of the most common parameters mapped in profile are air pollutant concentrations and sound levels. In each of those cases it may be important to analyze (air pollutant concentrations or sound levels) at varying heights so as to determine the air quality or noise health effects on people at different elevations, for example, living on different floor levels of an urban apartment. In actuality, both plan and profile view contour maps are used in air pollution and noise pollution studies.

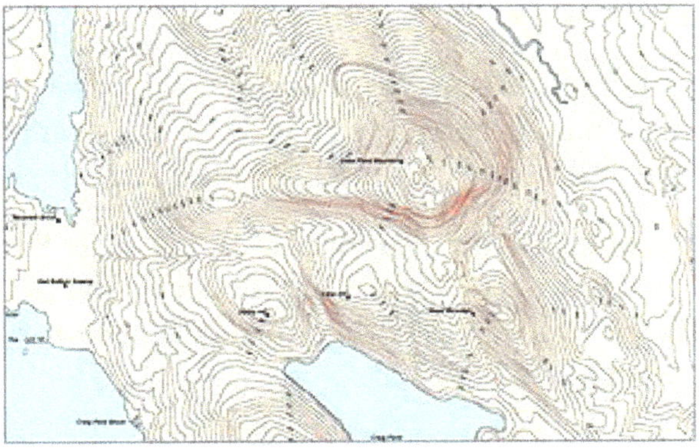

Contour map labeled aesthetically in an "elevation up" manner.

Labeling Contour Maps

Labels are a critical component of elevation maps. A properly labeled contour map helps the reader to quickly interpret the shape of the terrain. If numbers are placed close to each other, it means that the terrain is steep. Labels should be placed along a slightly curved line "pointing" to the summit or nadir, from several directions if possible, making the visual identification of the summit or nadir easy. Contour labels can be oriented so a reader is facing uphill when reading the label.

Manual labeling of contour maps is a time-consuming process, however, there are a few software systems that can do the job automatically and in accordance with cartographic conventions, called automatic label placement.

SCALE BARS

Scale bars provide a visual indication of the size of features, and distance between features, on the map. A scale bar is a line or bar divided into parts. It is labeled with its ground length, usually in multiples of map units, such as tens of kilometers or hundreds of miles. When a scale bar is added to the layout, it is associated with a map frame and maintains a connection to the map inside the frame. If the map scale changes, the scale bar updates to remain correct.

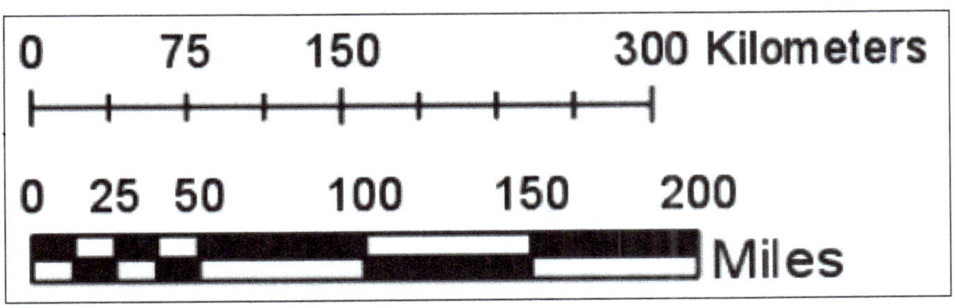

Insert a Scale Bar

When a scale bar is added to the page, it is automatically associated with the default map frame. If there isn't a map frame on the page, the map frame defaults to <None>. This property is not final and can be modified any time after the scale bar is inserted.

On the Insert tab, in the Map Surrounds group, click the Scale Bar button. Clicking the top half of the button inserts a default scale bar, while clicking the bottom half displays a gallery of all the scale bar style items in the project. Choose a style, and on the layout view, click and drag to create the scale bar.

Modify a Scale Bar

You can modify a scale bar's properties—such as its appearance, size, and position—using the Format tab on the ribbon and the Format Scale Bar pane.

Right-click the scale bar in the Contents pane and choose Properties to open the Format Scale Bar pane and set additional properties for the scale bar.

On the Options tab, you can rename the scale bar, switch its visibility on or off, or lock it so it cannot be selected on the layout.

You can also change the map frame that the scale bar references if you have multiple map frames on your layout. The Map Frame drop-down menu shows the currently selected map frame and all other map frames on the layout.

Map scales are often averages and can vary based on coordinate system, latitude, direction, and map extent. The Compute at center option calculates the map scale using the center of the map frame. This scale can differ from the scale in the map view, which, in many cases, calculates the scale at a location along the equator.

You can change the value for the Units, Label Text, Label Position (Before, After, Above, or Below), and Offset options of a scale bar, and you can customize its symbol using the Symbol option.

The Units group contains controls to change the units the scale bar uses as well as how they're displayed. Units sets the units of the scale bar, Label Position sets the position of the units label relative to the scale bar, and Label controls how the unit value is displayed.

In the Divisions group, you can set division and division values for the scale bar. Certain buttons may be unavailable depending on the current resize behavior. Division Value sets the value of one division, Divisions sets the number of divisions, Subdivisions sets the number of subdivisions, and checking the Show one division before zero check box updates the scale bar to show one division before the zero mark.

The scale bar Division Value option, as well as the behavior of the scale bar when it's resized or when the map scale changes, can be modified by choosing one of the following options from the Resize Behavior drop-down list:

- Adjust division value—Preserves the number of divisions and tries to preserve the scale bar width by adjusting the division value.
- Adjust number of divisions—Preserves the division value and tries to preserve the scale bar width by adjusting the number of divisions.
- Adjust divisions and division value—Preserves the number of subdivisions and tries to preserve the scale bar width by adjusting the number of divisions first, then the division values.
- Adjust width—Preserves the division value and number of divisions, and adjusts the scale bar width if the map scale changes.

You can also change the referenced scale bar in the Style group by clicking the drop-down menu.

Scale Bar Properties

The Properties tab allows you to modify the Numbers, Marks, and Bar settings that make up a scale bar.

The Numbers group contains controls to set the number display properties. Frequency sets the amount of numbers and their location. Position sets their display position relative to the bar. Additional options for number formatting can be set for Rounding, Use fractional characters, Show thousands separator, and Pad with Zeros. Pad with Zeros adds zeros after the decimal to match the Rounding value. Symbol can also be set.

The Marks group contains controls to set the tick mark properties. Frequency sets the amount of ticks and their location. Position sets their display position relative to the bar. Length and Symbol can also be set.

The Bar group allows you to modify the Height option of the bar by changing its increment, and you can modify the Symbol option.

For line-based scale bars, check the Stepped check box and adjust the Height and Symbol options to make them stepped scale bars. Stepped scale bars allow you to more clearly divisions and subdivisions.

Display

To modify the border, background, or shadow of the scale bar, click the Display tab. Each symbol's color and size can be set. The symbol's Rounding and X & Y gap options can also be modified.

COMPASS ROSE

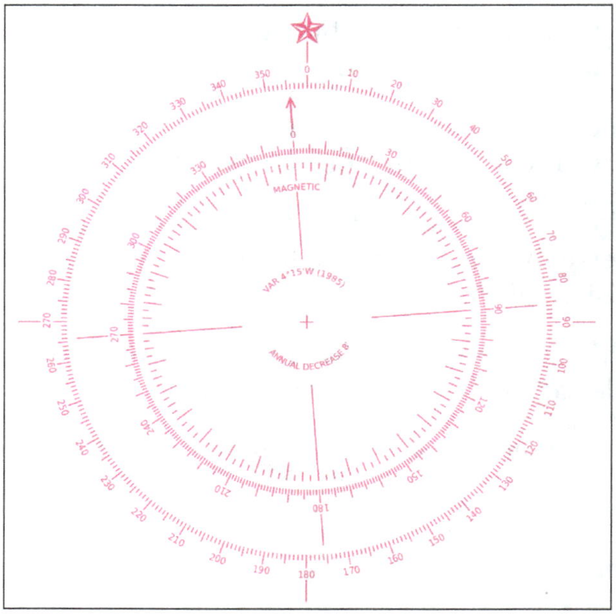

A common compass rose as found on a nautical chart showing both true north and magnetic north with magnetic variation. Also notice the correspondence between the 32-point rose (inner circle) and the modern 0–360° graduations.

A compass rose, sometimes called a windrose or Rose of the Winds, is a figure on a compass, map, nautical chart, or monument used to display the orientation of the cardinal directions (north, east, south, and west) and their intermediate points. It is also the term for the graduated markings found on the traditional magnetic compass. Today, the idea of a compass rose is found on, or featured in, almost all navigation systems, including nautical charts, non-directional beacons(NDB), VHF omnidirectional range (VOR) systems, global-positioning systems (GPS), and similar equipment.

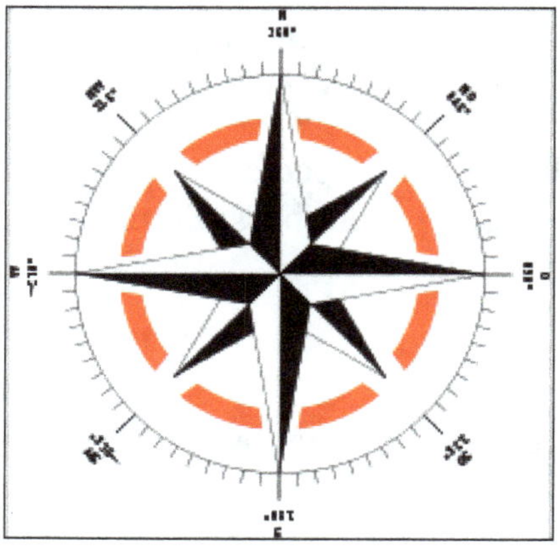

Compass rose with the eight principal winds.

The modern compass rose has eight principal winds. Listed clockwise, these are:

Compass point	Abbr.	Heading	Traditional wind
North	N	0°	Tramontane
North-East	NE	45° (45°×1)	Greco or Grecale
East	E	90° (45°×2)	Levante
South-East	SE	135° (45°×3)	Sirocco
South	S	180° (45°×4)	Ostro or Mezzogiorno
South-West	SW	225° (45°×5)	Libeccio or Garbino
West	W	270° (45°×6)	Ponente
North-West	NW	315° (45°×7)	Maestro or Mistral

Although modern compasses use the names of the eight principal directions (N, NE, E, SE, etc.), older compasses use the traditional Italianate wind names of Medieval origin (Tramontana, Greco, Levante, etc.)

4-point compass roses use only the four "basic winds" or "cardinal directions" (North, East, South, West), with angles of difference at 90°.

8-point compass roses use the eight principal winds—that is, the four cardinal directions (N, E, S, W) plus the four "intercardinal" or "ordinal directions" (NE, SE, SW, NW), at angles of difference of 45°.

16-point compass roses are constructed by bisecting the angles of the principal winds to come up with intermediate compass points, known as half-winds, at angles of difference of $22\frac{1}{2}°$. The names of the half-winds are simply combinations of the principal winds to either side, principal then ordinal. E.g. North-northeast (NNE), East-northeast (ENE), etc.

32-point compass roses are constructed by bisecting these angles, and coming up with quarter-winds at $11\frac{1}{4}°$ angles of difference. Quarter-wind names are constructed with the names "X by Y", which can be read as "one quarter wind from X toward Y", where X is one of the eight principal winds and Y is one of the two adjacent cardinal directions. For example, North-by-east (NbE) is one quarter wind from North towards East, North-east-by-north (NEbN) is one quarter wind from Northeast toward North. Naming all 32 points on the rose is called "boxing the compass".

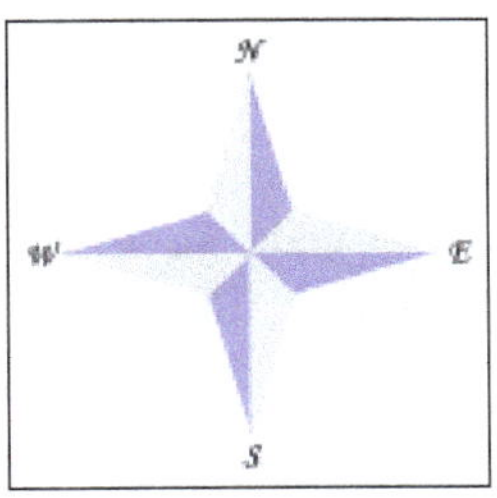

A 4-point compass rose.

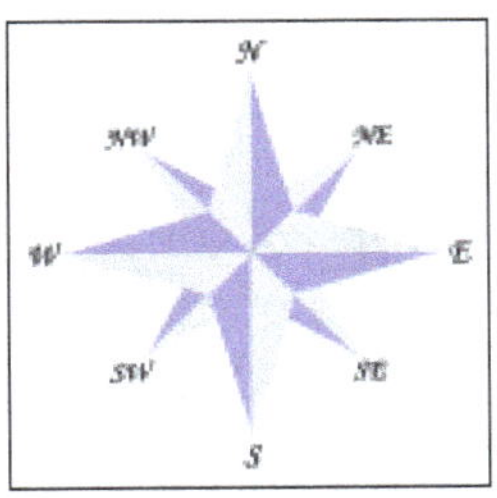

An 8-point compass rose.

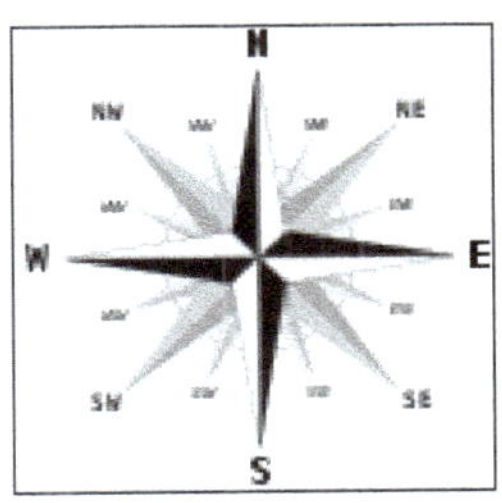

A 16-point compass rose.

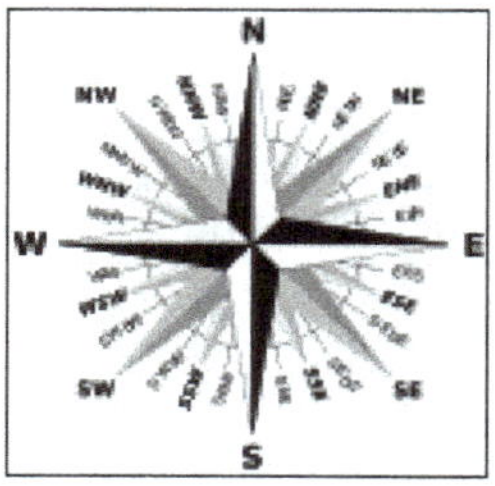

A 32-point compass rose.

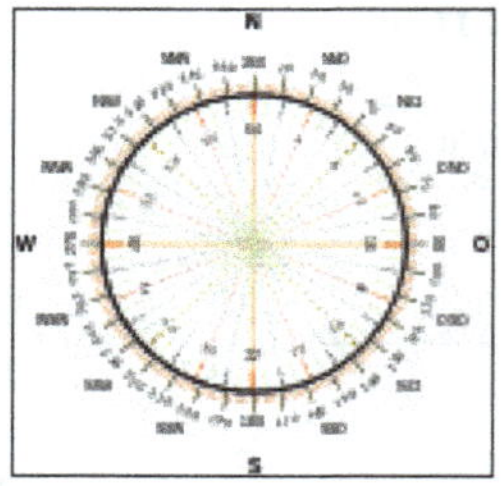

A 360 degree and 6400 NATO mil compass rose.

The 32-point rose has the uncomfortable number of $11\frac{1}{4}^\circ$ between points, but is easily found by halving divisions and may have been easier for those not using a 360° circle. Using gradians, of which there are 400 in a circle, the sixteen-point rose will have twenty-five gradians per point.

Linguistic anthropological studies have shown that most human communities have four points of cardinal direction. The names given to these directions are usually derived from either locally-specific geographic features (e.g. "towards the hills", "towards the sea") or from celestial bodies (especially the sun) or from atmospheric features (winds, temperature). Most mobile populations tend to adopt sunrise and sunset for East and West and the direction from where different winds blow to denote North and South.

Classical Compass Rose

The ancient Greeks originally maintained distinct and separate systems of points and winds. The four Greek cardinal points (arctos, anatole, mesembria, and dusis) were based on celestial bodies and used for orientation. The four Greek winds (Boreas, Notos, Eurus, Zephyrus) were confined to meteorology. Nonetheless, both systems were gradually conflated, and wind names came to eventually denote cardinal directions as well.

In his meteorological studies, Aristotle identified ten distinct winds: two north-south winds (Aparctias, Notos) and four sets of east-west winds blowing from different latitudes—the Arctic circle (Meses, Thrascias), the summer solstice horizon (Caecias, Argestes), the equinox (Apeliotes, Zephyrus) and the winter solstice (Eurus, Lips). However, Aristotle's system was asymmetric. To restore balance, Timosthenes of Rhodes added two more winds to produce the classical 12-wind rose, and began using the winds to denote geographical direction in navigation. Eratosthenes deducted two winds from Aristotle's system, to produce the classical 8-wind rose.

The Romans (e.g. Seneca, Pliny) adopted the Greek 12-wind system, and replaced its names with Latin equivalents, e.g. Septentrio, Subsolanus, Auster, Favonius, etc. Uniquely, Vitruvius came up with a 24-wind rose.

According to the chronicler Einhard, the Frankish king Charlemagne himself came up with his own names for the classical 12 winds. Intermediate winds were constructed as simple compound names of these four (e.g. "Nordostdroni", the "northeasterly" wind). These Carolingian names are the source of the modern compass point names found in nearly all modern west European languages. (e.g. North, East, South and West in English; Nord, Est, Sud, Ouest in French, etc.)

The following table gives a rough equivalence of the classical 12-wind rose with the modern compass directions (The directions are imprecise since it is not clear at what angles the classical winds are supposed to be with each other; some have argued that they should be equally spaced at 30 degrees each.)

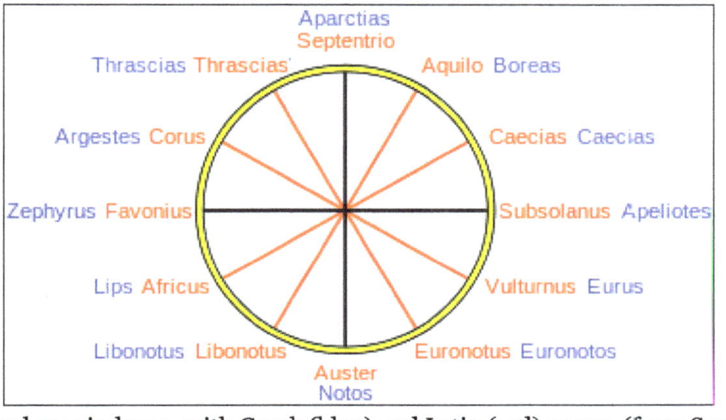

Classical 12-wind rose, with Greek (blue) and Latin (red) names (from Seneca).

Sidereal Compass Rose

The "sidereal" compass rose demarcates the compass points by the position of stars in the night sky, rather than winds. Arab navigators in the Red Sea and the Indian Ocean, who depended on celestial navigation, were using a 32-point sidereal compass rose before the end of the 10th century. In the northern hemisphere, the steady Pole Star (Polaris) was used for the N-S axis; the less-steady Southern Cross had to do for the southern hemisphere, as the southern pole star, Sigma Octantis, is too dim to be easily seen from Earth with the naked eye. The other thirty points on the sidereal rose were determined by the rising and setting positions of fifteen bright stars. Reading from North to South, in their rising and setting positions, these are:

Point	Star
N	Polaris
NbE	"the Guards" (Ursa Minor)
NNE	Alpha Ursa Major
NEbN	Alpha Cassiopeiae
NE	Capella
NEbE	Vega

ENE	Arcturus
EbN	the Pleiades
E	Altair
EbS	Orion's belt
ESE	Sirius
SEbE	Beta Scorpionis
SE	Antares
SEbS	Alpha Centauri
SSE	Canopus
SbE	Achernar
S	Southern Cross

The western half of the rose would be the same stars in their setting position. The true position of these stars is only approximate to their theoretical equidistant rhumbs on the sidereal compass. Stars with the same declination formed a "linear constellation" or *kavenga* to provide direction as the night progressed.

A similar sidereal compass was used by Polynesian and Micronesian navigators in the Pacific Ocean, although different stars were used in a number of cases, clustering around the East-West axis.

Mariner's Compass Rose

In Europe, the Classical 12-wind system continued to be taught in academic settings during the Medieval era, but seafarers in the Mediterranean came up with their own distinct 8-wind system. The mariners used names derived from the Mediterranean lingua franca—the Italian-tinged patois among Medieval sailors, composed principally of Ligurian, mixed with Venetian, Sicilian, Provençal, Catalan, Greek and Arabic terms from around the Mediterranean basin.

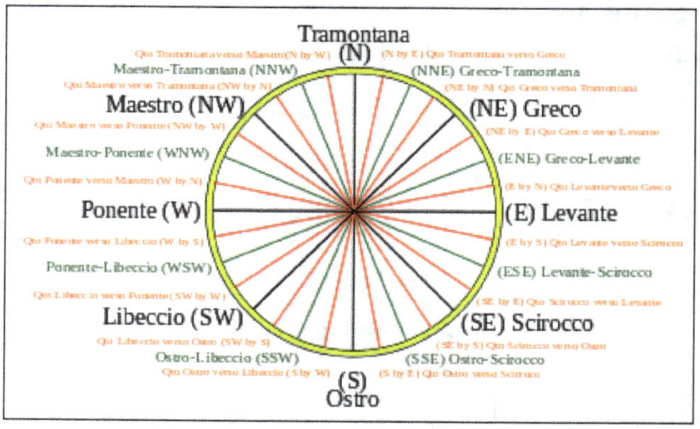

32-wind compass with traditional names (and traditional color code).

- (N) Tramontana,
- (NE) Greco (or Bora),
- (E) Levante,
- (SE) Scirocco (or Exaloc),
- (S) Ostro (or Mezzogiorno),
- (SW) Libeccio (or Garbino),
- (W) Ponente,
- (NW) Maestro (or Mistral).

The exact origin of the mariner's eight-wind rose is obscure. Only two of its point names (*Ostro*, *Libeccio*) have Classical etymologies, the rest of the names seem to be autonomously derived. Two Arabic words stand out: *Scirocco* (SE) from *al-Sharq* and the variant *Garbino* (SW), from *al-Gharb*. This suggests the mariner's rose was probably acquired by southern Italian seafarers not from their classical Roman ancestors, but rather from Norman Sicily in the 11th to 12th centuries. The coasts of the Maghreb and Mashriqare SW and SE of Sicily respectively; the *Greco* (a NE wind), reflects the position of Byzantine-held Calabria-Apulia to the northeast of Arab Sicily, while the *Maestro* (a NW wind) is a reference to the Mistral wind that blows from the southern French coast towards northwest Sicily.

The 32-point compass used for navigation in the Mediterranean by the 14th century, had increments of $11\frac{1}{4}°$ between points. Only the eight principal winds (N, NE, E, SE, S, SW, W, NW) were given special names. The eight half-winds just combined the names of the two principal winds, e.g. Greco-Tramontana for NNE, Greco-Levante for ENE, and so on. Quarter-winds were more cumbersomely phrased, with the closest principal wind named first and the next-closest principal wind second, e.g. "Quarto di Tramontana verso Greco" (literally, "one quarter wind from North towards Northeast", i.e. North by East), and "Quarto di Greco verso Tramontana" ("one quarter wind from NE towards N", i.e. Northeast by North). Boxing the compass (naming all 32 winds) was expected of all Medieval mariners.

Depiction on Nautical Charts

In the earliest Medieval portolan charts of the 14th century, compass roses were depicted as mere collections of color-coded compass rhumb lines: black for the eight main winds, green for the eight half-winds and red for the sixteen quarter-winds. The average portolan chart had sixteen such roses (or confluence of lines), spaced out equally around the circumference of a large implicit circle.

The cartographer Cresques Abraham of Majorca, in his Catalan Atlas of 1375, was the first to draw an ornate compass rose on a map. By the end of the 15th century, Portuguese cartographers began drawing multiple ornate compass roses throughout the chart, one upon each of the sixteen circumference roses (unless the illustration conflicted with coastal details).

The points on a compass rose were frequently labeled by the initial letters of the mariner's principal winds (T, G, L, S, O, L, P, M). However, from the outset, the custom also began to distinguish the north from the other points by a specific visual marker. Medieval Italian cartographers typically used a simple arrowhead or circumflex-hatted T (an allusion to the compass needle) to designate the north, while the Majorcan cartographic school typically used a stylized Pole Star for its north mark. The use of the fleur-de-lis as north mark was introduced by Pedro Reinel, and quickly became customary in compass roses (and is still often used today). Old compass roses also often used a Christian cross at Levante (E), indicating the direction of Jerusalem from the point of view of the Mediterranean sea.

The twelve Classical winds (or a subset of them) were also sometimes depicted on portolan charts, albeit not on a compass rose, but rather separately on small disks or coins on the edges of the map.

The compass rose was also depicted on traverse boards used on board ships to record headings sailed at set time intervals.

Early 32-wind compass rose, shown as a mere collection of color-coded rhumblines, from a Genoese nautical chart.

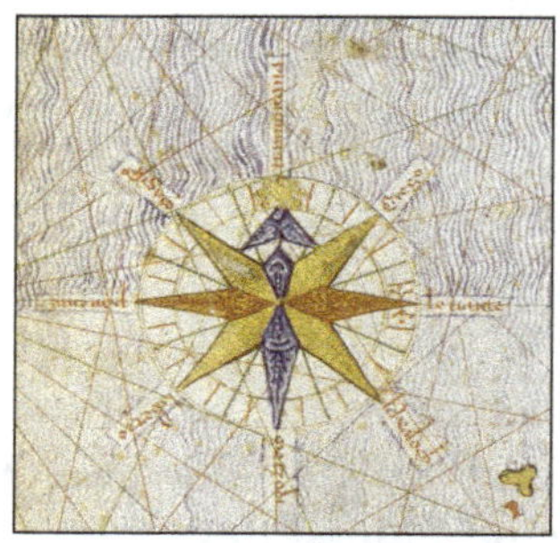

First ornate compass rose depicted on a chart, from the Catalan Atlas,with the Pole Star as north mark.

More ornate compass rose, with letters of traditional winds and compass needle as north mark, from a nautical chart by Jorge de Aguiar.

Highly ornate compass rose, with fleur-de-lis as north mark and cross pattée as east mark, from the Cantino planisphere.

Modern Depictions

A 16-point compass rose on the grounds of a library serves both as a pedagogical device and public art.

The contemporary compass rose appears as two rings, one smaller and set inside the other. The outside ring denotes truecardinal directions while the smaller inside ring denotes magnetic cardinal directions. True north refers to the geographical location of the north pole while magnetic north refers to the direction towards which the north pole of a magnetic object (as found in a compass) will point. The angular difference between true and magnetic north is called variation, which varies depending on location. The angular difference between magnetic heading and compass heading is called deviation which varies by vessel and its heading.

Use as Symbol

- The NATO symbol uses a four pointed rose.
- Outward Bound uses the compass rose as the logo for various schools around the world.
- An 8-point compass rose was the logo of Varig, the largest airline in Brazil for many decades until its bankruptcy in 2006.
- An 8-point compass rose is a prominent feature in the logo of the Seattle Mariners Major League Baseball club.
- Hong Kong Correctional Services's crest uses four point star.
- The compass rose is used as the symbol of the worldwide Anglican Communion of churches.
- A 16-point compass rose was IBM's logo for the System/360 product line.

- A 16-point compass rose is the official logo of the Spanish National University of Distance Education (Universidad Nacional de Educación a Distancia or UNED).
- A 16-point compass rose is present on the seal and the flag of the Central Intelligence Agency of the Federal government of the United States (the CIA).

MAP PROJECTION

A medieval depiction of the Ecumene, constructed after the coordinates in Ptolemy's *Geography* and using his second map projection.

A map projection is a systematic transformation of the latitudes and longitudes of locations from the surface of a sphere or an ellipsoid into locations on a plane. Maps cannot be created without map projections. All map projections necessarily distort the surface in some fashion. There is no limit to the number of possible map projections.

Projections are a subject of several pure mathematical fields, including differential geometry, projective geometry, and manifolds. However, "map projection" refers specifically to a cartographic projection.

Depending on the purpose of the map, some distortions are acceptable and others are not; therefore, different map projections exist in order to preserve some properties of the sphere-like body at the expense of other properties.

The surfaces of planetary bodies can be mapped even if they are too irregular to be modeled well with a sphere or ellipsoid.

The surface to be mapped is that of a sphere. The Earth and other large celestial bodies are generally better modeled as oblate spheroids, whereas small objects such as asteroids often have irregular shapes. Io is better modeled by triaxial ellipsoid or prolated spheroid with small eccentricities. Haumea's shape is a Jacobi ellipsoid, with its major axis twice as long as its minor and with its middle axis one and half times as long as its minor. These other surfaces can be mapped as well. Therefore, more generally, a map projection is any method of "flattening" a continuous curved surface onto a plane.

Carl Friedrich Gauss's Theorema Egregium proved that a sphere's surface cannot be represented on a plane without distortion. The same applies to other reference surfaces used as models for the Earth, such as oblate spheroids, ellipsoids and geoids. Since any map projection is a representation of one of those surfaces on a plane, all map projections distort. Every distinct map projection distorts in a distinct way. The study of map projections is the characterization of these distortions.

Maps can be more useful than globes in many situations: they are more compact and easier to store; they readily accommodate an enormous range of scales; they are viewed easily on computer displays; they can facilitate measuring properties of the region being mapped; they can show larger portions of the Earth's surface at once; and they are cheaper to produce and transport. These useful traits of maps motivate the development of map projections.

Projection is not limited to perspective projections, such as those resulting from casting a shadow on a screen, or the rectilinear image produced by a pinhole camera on a flat film plate. Rather, any mathematical function transforming coordinates from the curved surface to the plane is a projection. Few projections in actual use are perspective.

Metric Properties of Maps

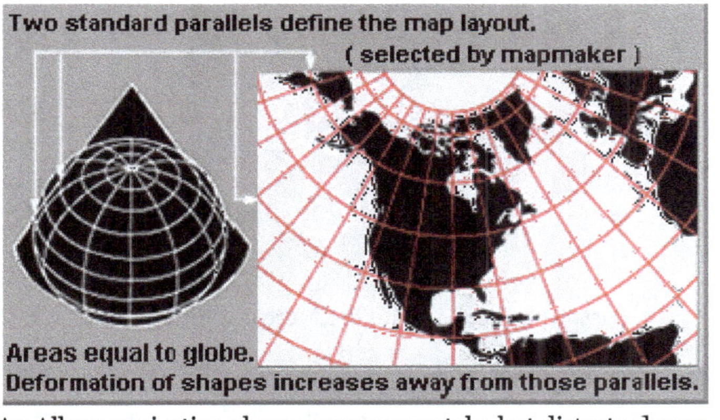

An Albers projection shows areas accurately, but distorts shapes.

Many properties can be measured on the Earth's surface independent of its geography. Some of these properties are:

- Area,
- Shape,
- Direction,
- Bearing,

- Distance,
- Scale.

Map projections can be constructed to preserve at least one of these properties, though only in a limited way for most. Each projection preserves, compromises, or approximates basic metric properties in different ways. The purpose of the map determines which projection should form the base for the map. Because many purposes exist for maps, a diversity of projections have been created to suit those purposes.

Another consideration in the configuration of a projection is its compatibility with data sets to be used on the map. Data sets are geographic information; their collection depends on the chosen datum (model) of the Earth. Different datums assign slightly different coordinates to the same location, so in large scale maps, such as those from national mapping systems, it is important to match the datum to the projection. The slight differences in coordinate assignation between different datums is not a concern for world maps or other vast territories, where such differences get shrunk to imperceptibility.

Distortion

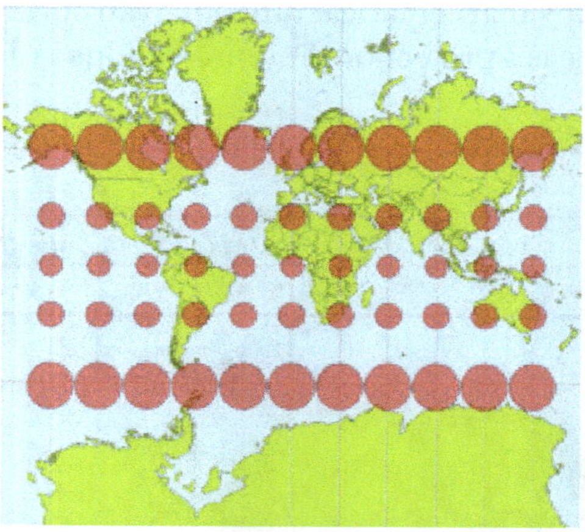

Tissot's Indicatrices on the Mercator projection.

The classical way of showing the distortion inherent in a projection is to use Tissot's indicatrix. For a given point, using the scale factor h along the meridian, the scale factor k along the parallel, and the angle θ' between them, Nicolas Tissot described how to construct an ellipse that characterizes the amount and orientation of the components of distortion. By spacing the ellipses regularly along the meridians and parallels, the network of indicatrices shows how distortion varies across the map.

Design and Construction

The creation of a map projection involves two steps:

1. Selection of a model for the shape of the Earth or planetary body (usually

choosing between a sphere or ellipsoid). Because the Earth's actual shape is irregular, information is lost in this step.

2. Transformation of geographic coordinates (longitude and latitude) to Cartesian (x,y) or polar plane coordinates. In large-scale maps, Cartesian coordinates normally have a simple relation to eastings and northings defined as a grid superimposed on the projection. In small-scale maps, eastings and northings are not meaningful, and grids are not superimposed.

Some of the simplest map projections are literal projections, as obtained by placing a light source at some definite point relative to the globe and projecting its features onto a specified surface. This is not the case for most projections, which are defined only in terms of mathematical formulae that have no direct geometric interpretation. However, picturing the light source-globe model can be helpful in understanding the basic concept of a map projection.

Choosing a Projection Surface

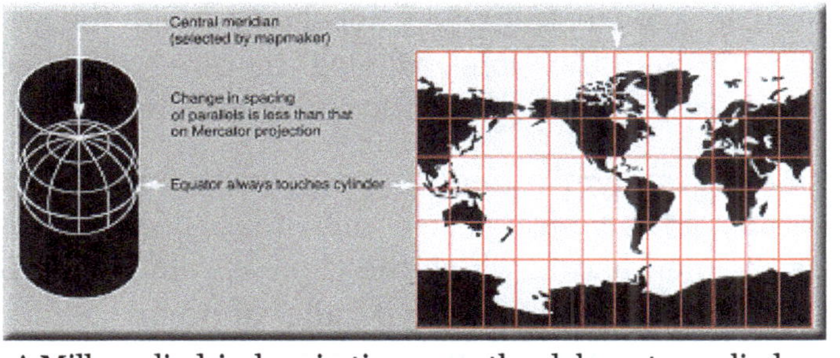

A Miller cylindrical projection maps the globe onto a cylinder.

A surface that can be unfolded or unrolled into a plane or sheet without stretching, tearing or shrinking is called a *developable surface*. The cylinder, cone and the plane are all developable surfaces. The sphere and ellipsoid do not have developable surfaces, so any projection of them onto a plane will have to distort the image. (To compare, one cannot flatten an orange peel without tearing and warping it.)

One way of describing a projection is first to project from the Earth's surface to a developable surface such as a cylinder or cone, and then to unroll the surface into a plane. While the first step inevitably distorts some properties of the globe, the developable surface can then be unfolded without further distortion.

Aspect of the Projection

Once a choice is made between projecting onto a cylinder, cone, or plane, the aspect of the shape must be specified. The aspect describes how the developable surface is placed relative to the globe: it may be *normal* (such that the surface's axis of symmetry coincides with the Earth's axis), *transverse* (at right angles to the Earth's axis) or *oblique* (any angle in between).

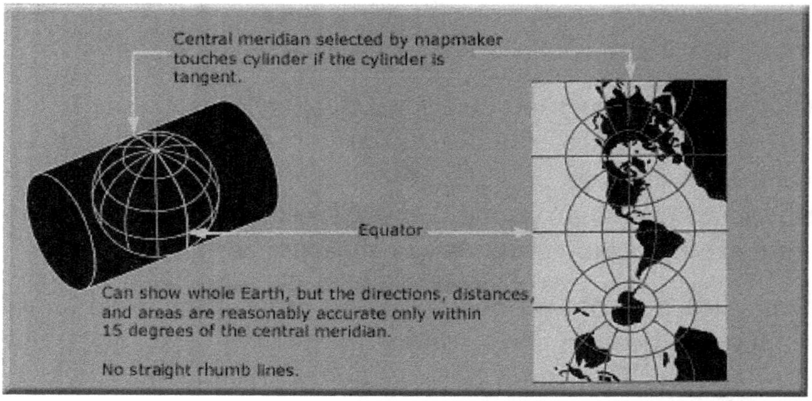

This transverse Mercator projection is mathematically the same as a standard Mercator, but oriented around a different axis.

Notable Lines

The developable surface may also be either *tangent* or *secant* to the sphere or ellipsoid. Tangent means the surface touches but does not slice through the globe; secant means the surface does slice through the globe. Moving the developable surface away from contact with the globe never preserves or optimizes metric properties.

Tangent and secant lines (*standard lines*) are represented undistorted. If these lines are a parallel of latitude, as in conical projections, it is called a *standard parallel*. The *central meridian* is the meridian to which the globe is rotated before projecting. The central meridian (usually written λ_0) and a parallel of origin (usually written φ_0) are often used to define the origin of the map projection.

Scale

A globe is the only way to represent the earth with constant scale throughout the entire map in all directions. A map cannot achieve that property for any area, no matter how small. It can, however, achieve constant scale along specific lines.

Some possible properties are:

- The scale depends on location, but not on direction. This is equivalent to preservation of angles, the defining characteristic of a conformal map.
- Scale is constant along any parallel in the direction of the parallel. This applies for any cylindrical or pseudocylindrical projection in normal aspect.
- Combination of the above: the scale depends on latitude only, not on longitude or direction. This applies for the Mercator projection in normal aspect.
- Scale is constant along all straight lines radiating from a particular geographic location. This is the defining characteristic of an equidistant projection such

as the Azimuthal equidistant projection. There are also projections (Maurer's Two-point equidistant projection, Close) where true distances from *two* points are preserved.

Choosing a Model for the Shape of the Body

Projection construction is also affected by how the shape of the Earth or planetary body is approximated. However, the Earth's actual shape is closer to an oblate ellipsoid. Whether spherical or ellipsoidal, the principles discussed hold without loss of generality.

Selecting a model for a shape of the Earth involves choosing between the advantages and disadvantages of a sphere versus an ellipsoid. Spherical models are useful for small-scale maps such as world atlases and globes, since the error at that scale is not usually noticeable or important enough to justify using the more complicated ellipsoid. The ellipsoidal model is commonly used to construct topographic maps and for other large and medium-scale maps that need to accurately depict the land surface. Auxiliary latitudes are often employed in projecting the ellipsoid.

A third model is the geoid, a more complex and accurate representation of Earth's shape coincident with what mean sea level would be if there were no winds, tides, or land. Compared to the best fitting ellipsoid, a geoidal model would change the characterization of important properties such as distance, conformality and equivalence. Therefore, in geoidal projections that preserve such properties, the mapped graticule would deviate from a mapped ellipsoid's graticule. Normally the geoid is not used as an Earth model for projections, however, because Earth's shape is very regular, with the undulation of the geoid amounting to less than 100 m from the ellipsoidal model out of the 6.3 million m Earth radius. For irregular planetary bodies such as asteroids, however, sometimes models analogous to the geoid are used to project maps from.

Classification

A fundamental projection classification is based on the type of projection surface onto which the globe is conceptually projected. The projections are described in terms of placing a gigantic surface in contact with the earth, followed by an implied scaling operation. These surfaces are cylindrical (e.g. Mercator), conic (e.g. Albers), and plane (e.g. stereographic). Many mathematical projections, however, do not neatly fit into any of these three conceptual projection methods. Hence other peer categories have been described in the literature, such as pseudoconic, pseudocylindrical, pseudoazimuthal, retroazimuthal, and polyconic.

Another way to classify projections is according to properties of the model they preserve. Some of the more common categories are:

- Preserving direction (*azimuthal or zenithal*), a trait possible only from one or two points to every other point.

- Preserving shape locally (*conformal* or *orthomorphic*).
- Preserving area (*equal-area* or *equiareal* or *equivalent* or *authalic*).
- Preserving distance (*equidistant*), a trait possible only between one or two points and every other point.
- Preserving shortest route, a trait preserved only by the gnomonic projection.

Because the sphere is not a developable surface, it is impossible to construct a map projection that is both equal-area and conformal.

Projections by Surface

The three developable surfaces (plane, cylinder, cone) provide useful models for understanding, describing, and developing map projections. However, these models are limited in two fundamental ways. For one thing, most world projections in use do not fall into any of those categories. For another thing, even most projections that do fall into those categories are not naturally attainable through physical projection. As L.P. Lee notes,

> "No reference has been made in the above definitions to cylinders, cones or planes. The projections are termed cylindric or conic because they can be regarded as developed on a cylinder or a cone, as the case may be, but it is as well to dispense with picturing cylinders and cones, since they have given rise to much misunderstanding. Particularly is this so with regard to the conic projections with two standard parallels: They may be regarded as developed on cones, but they are cones which bear no simple relationship to the sphere. In reality, cylinders and cones provide us with convenient descriptive terms, but little else".

Lee's objection refers to the way the terms *cylindrical*, *conic*, and *planar* (azimuthal) have been abstracted in the field of map projections. If maps were projected as in light shining through a globe onto a developable surface, then the spacing of parallels would follow a very limited set of possibilities. Such a cylindrical projection (for example) is one which:

- Is rectangular;
- Has straight vertical meridians, spaced evenly;
- Has straight parallels symmetrically placed about the equator;
- Has parallels constrained to where they fall when light shines through the globe onto the cylinder, with the light source someplace along the line formed by the intersection of the prime meridian with the equator, and the center of the sphere.

(If you rotate the globe before projecting then the parallels and meridians will not necessarily still be straight lines. Rotations are normally ignored for the purpose of classification.)

Where the light source emanates along the line described in this last constraint is what yields the differences between the various "natural" cylindrical projections. But the term *cylindrical* as used in the field of map projections relaxes the last constraint entirely. Instead the parallels can be placed according to any algorithm the designer has decided suits the needs of the map. The famous Mercator projection is one in which the placement of parallels does not arise by "projection"; instead parallels are placed how they need to be in order to satisfy the property that a course of constant bearing is always plotted as a straight line.

Cylindrical

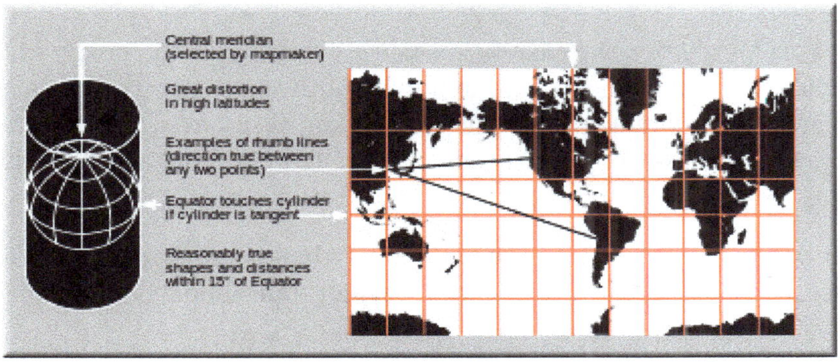

The Mercator projection shows rhumbs as straight lines. A rhumb is a course of constant bearing. Bearing is the compass direction of movement.

A "normal cylindrical projection" is any projection in which meridians are mapped to equally spaced vertical lines and circles of latitude (parallels) are mapped to horizontal lines.

The mapping of meridians to vertical lines can be visualized by imagining a cylinder whose axis coincides with the Earth's axis of rotation. This cylinder is wrapped around the Earth, projected onto, and then unrolled.

By the geometry of their construction, cylindrical projections stretch distances east-west. The amount of stretch is the same at any chosen latitude on all cylindrical projections, and is given by the secant of the latitude as a multiple of the equator's scale. The various cylindrical projections are distinguished from each other solely by their north-south stretching (where latitude is given by φ):

- North-south stretching equals east-west stretching (sec φ): The east-west scale matches the north-south scale: conformal cylindrical or Mercator; this distorts areas excessively in high latitudes.
- North-south stretching grows with latitude faster than east-west stretching ($\sec^2 \varphi$): The cylindric perspective (or central cylindrical) projection; unsuitable because distortion is even worse than in the Mercator projection.
- North-south stretching grows with latitude, but less quickly than the east-west stretching: such as the Miller cylindrical projection (sec _ φ).

- North-south distances neither stretched nor compressed (1): equirectangular projection or "plate carrée".
- North-south compression equals the cosine of the latitude (the reciprocal of east-west stretching): equal-area cylindrical. This projection has many named specializations differing only in the scaling constant, such as the Gall–Peters or Gall orthographic (undistorted at the 45° parallels), Behrmann (undistorted at the 30° parallels), and Lambert cylindrical equal-area (undistorted at the equator). Since this projection scales north-south distances by the reciprocal of east-west stretching, it preserves area at the expense of shapes.

In the first case (Mercator), the east-west scale always equals the north-south scale. In the second case (central cylindrical), the north-south scale exceeds the east-west scale everywhere away from the equator. Each remaining case has a pair of secant lines—a pair of identical latitudes of opposite sign (or else the equator) at which the east-west scale matches the north-south-scale.

Normal cylindrical projections map the whole Earth as a finite rectangle, except in the first two cases, where the rectangle stretches infinitely tall while retaining constant width.

Pseudocylindrical

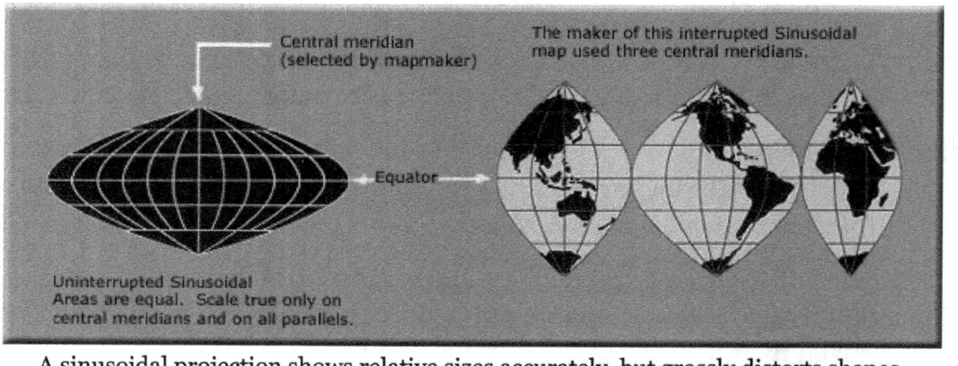

A sinusoidal projection shows relative sizes accurately, but grossly distorts shapes. Distortion can be reduced by "interrupting" the map.

Pseudocylindrical projections represent the *central* meridian as a straight line segment. Other meridians are longer than the central meridian and bow outward, away from the central meridian. Pseudocylindrical projections map parallels as straight lines. Along parallels, each point from the surface is mapped at a distance from the central meridian that is proportional to its difference in longitude from the central meridian. Therefore, meridians are equally spaced along a given parallel. On a pseudocylindrical map, any point further from the equator than some other point has a higher latitude than the other point, preserving north-south relationships. This trait is useful when illustrating phenomena that depend on latitude, such as climate. Examples of pseudocylindrical projections include:

- Sinusoidal, which was the first pseudocylindrical projection developed. On the map, as in reality, the length of each parallel is proportional to the cosine of the latitude. The area of any region is true.

- Collignon projection, which in its most common forms represents each meridian as two straight line segments, one from each pole to the equator.

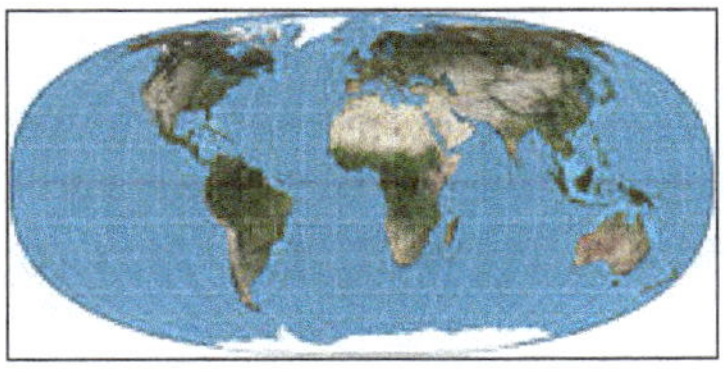

Tobler hyperelliptical

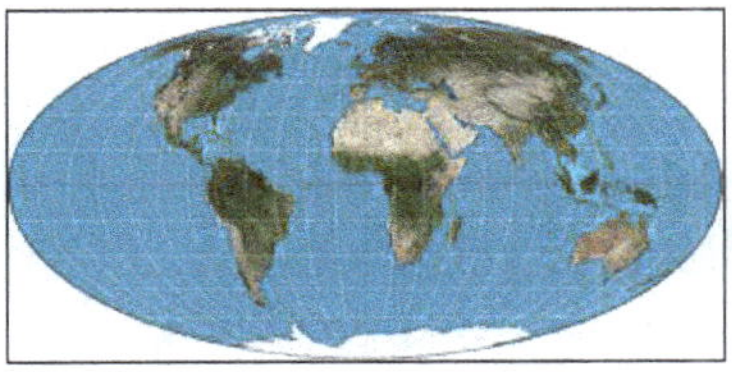

Mollweide

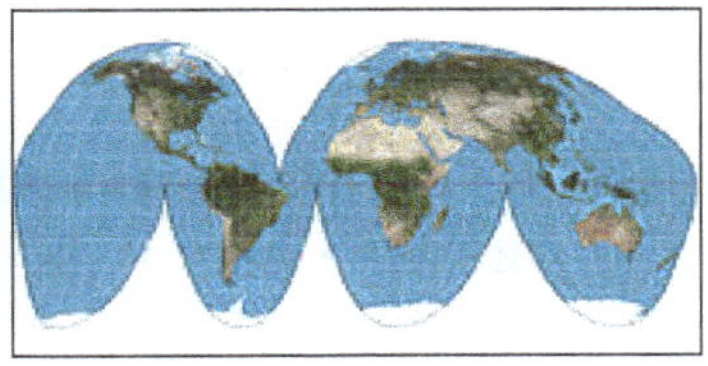

Goode homolosine

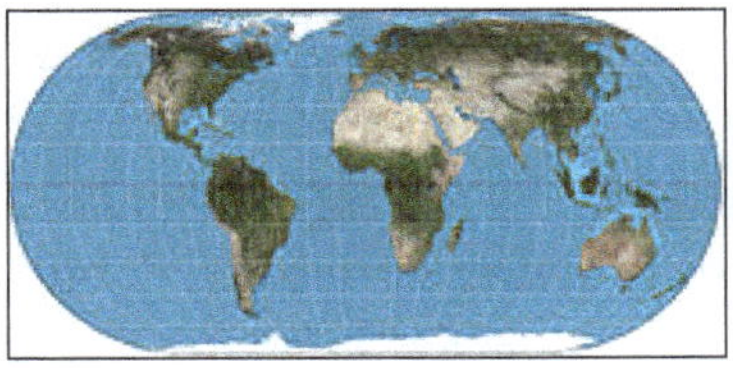

Eckert IV

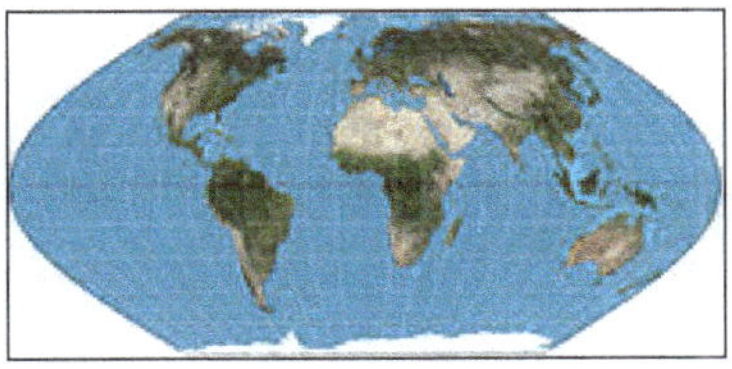

Eckert VI

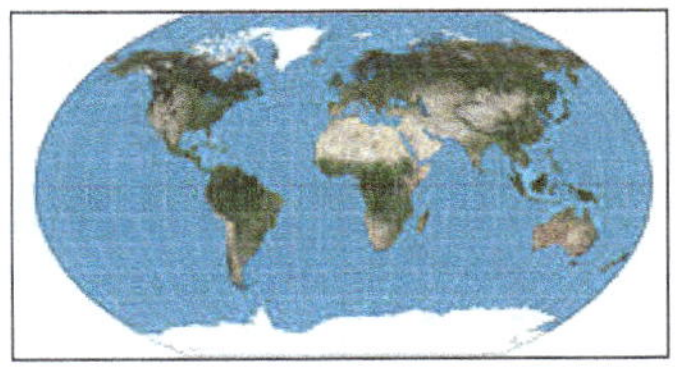

Kavrayskiy VII

Hybrid

The HEALPix projection combines an equal-area cylindrical projection in equatorial regions with the Collignon projection in polar areas.

Conic

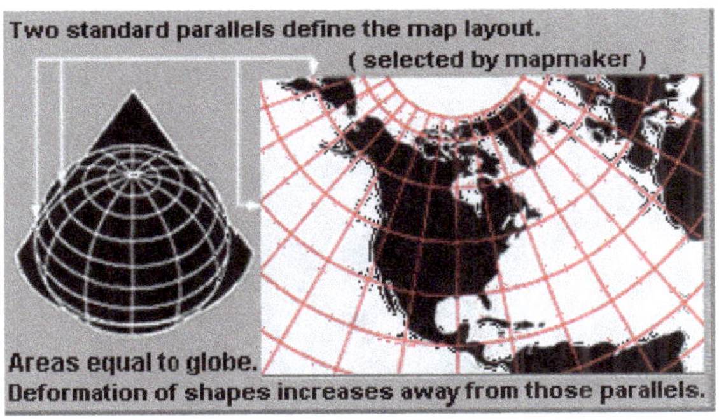

Albers conic.

The term "conic projection" is used to refer to any projection in which meridians are mapped to equally spaced lines radiating out from the apex and circles of latitude (parallels) are mapped to circular arcs centered on the apex.

When making a conic map, the map maker arbitrarily picks two standard parallels. Those standard parallels may be visualized as secant lines where the cone intersects the globe—or, if the map maker chooses the same parallel twice, as the tangent line where the cone is tangent to the globe. The resulting conic map has low distortion in scale, shape, and area near those standard parallels. Distances along the parallels to the north of both standard parallels or to the south of both standard parallels are stretched; distances along parallels between the standard parallels are compressed. When a single standard parallel is used, distances along all other parallels are stretched.

Conic projections that are commonly used are:

- Equidistant conic, which keeps parallels evenly spaced along the meridians to preserve a constant distance scale along each meridian, typically the same or similar scale as along the standard parallels.
- Albers conic, which adjusts the north-south distance between non-standard parallels to compensate for the east-west stretching or compression, giving an equal-area map.
- Lambert conformal conic, which adjusts the north-south distance between non-standard parallels to equal the east-west stretching, giving a conformal map.

Pseudoconic

- Bonne, an equal-area projection on which most meridians and parallels appear as curved lines. It has a configurable standard parallel along which there is no distortion.
- Werner cordiform, upon which distances are correct from one pole, as well as along all parallels.
- American polyconic.

Azimuthal (Projections onto a Plane)

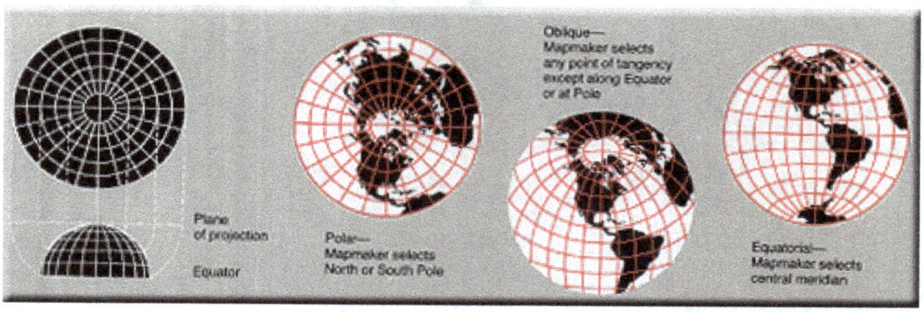

An azimuthal equidistant projection shows distances and directions accurately from the center point, but distorts shapes and sizes elsewhere.

Azimuthal projections have the property that directions from a central point are preserved and therefore great circles through the central point are represented by straight lines on the map. These projections also have radial symmetry in the scales and hence in the distortions: map distances from the central point are computed by a function $r(d)$ of the true distance d, independent of the angle; correspondingly, circles with the central point as center are mapped into circles which have as center the central point on the map.

The mapping of radial lines can be visualized by imagining a plane tangent to the Earth, with the central point as tangent point.

The radial scale is $r'(d)$ and the transverse scale $r(d)/(R \sin \frac{d}{R})$ where R is the radius of the Earth.

Some azimuthal projections are true perspective projections; that is, they can be constructed mechanically, projecting the surface of the Earth by extending lines from a point of perspective (along an infinite line through the tangent point and the tangent point's antipode) onto the plane:

- The gnomonic projection displays great circles as straight lines. Can be constructed by using a point of perspective at the center of the Earth. $r(d) = c \tan \frac{d}{R}$; so that even just a hemisphere is already infinite in extent.
- The General Perspective projection can be constructed by using a point of perspective outside the earth. Photographs of Earth (such as those from the International Space Station) give this perspective.
- The orthographic projection maps each point on the earth to the closest point on the plane. Can be constructed from a point of perspective an infinite distance from the tangent point; $r(d) = c \sin \frac{d}{R}$. Can display up to a hemisphere on a finite circle. Photographs of Earth from far enough away, such as the Moon, approximate this perspective.
- The stereographic projection, which is conformal, can be constructed by using the tangent point's antipode as the point of perspective. $r(d) = c \tan \frac{d}{2R}$; the scale is $c/(2R \cos^2 \frac{d}{2R})$. Can display nearly the entire sphere's surface on a finite circle. The sphere's full surface requires an infinite map.

Other azimuthal projections are not true perspective projections:

- Azimuthal equidistant: $r(d) = cd$; it is used by amateur radio operators to know the direction to point their antennas toward a point and see the distance to it. Distance from the tangent point on the map is proportional to surface distance on the earth.
- Lambert azimuthal equal-area. Distance from the tangent point on the map is proportional to straight-line distance through the earth: $r(d) = c \sin \frac{d}{2R}$.

- Logarithmic azimuthal is constructed so that each point's distance from the center of the map is the logarithm of its distance from the tangent point on the Earth. $r(d) = c \ln \frac{d}{d_0}$); locations closer than at a distance equal to the constant d_0 are not shown.

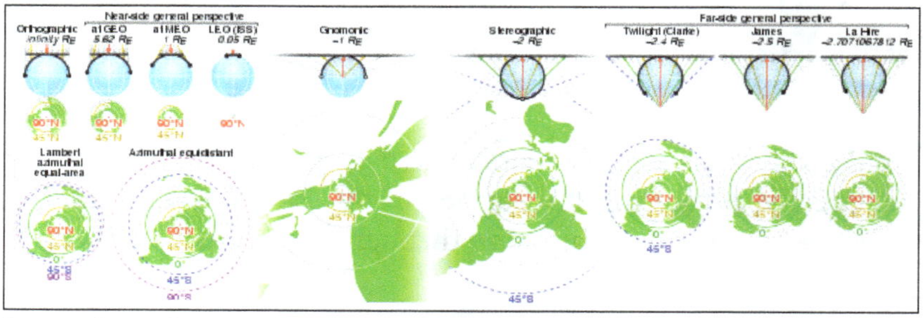

Comparison of some azimuthal projections centred on 90° N at the same scale, ordered by projection altitude in Earth radii.

Projections by Preservation of a Metric Property

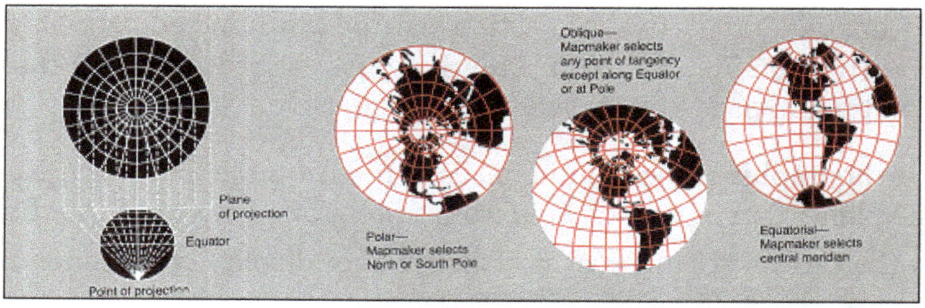

A stereographic projection is conformal and perspective but not equal area or equidistant.

Conformal

Conformal, or orthomorphic, map projections preserve angles locally, implying that they map infinitesimal circles of constant size anywhere on the Earth to infinitesimal circles of varying sizes on the map. In contrast, mappings that are not conformal distort most such small circles into ellipses of distortion. An important consequence of conformality is that relative angles at each point of the map are correct, and the local scale (although varying throughout the map) in every direction around any one point is constant. These are some conformal projections:

- Mercator: Rhumb lines are represented by straight segments,
- Transverse Mercator,
- Stereographic: Any circle of a sphere, great and small, maps to a circle or straight line,
- Roussilhe,

- Lambert conformal conic,
- Peirce quincuncial projection,
- Adams hemisphere-in-a-square projection,
- Guyou hemisphere-in-a-square projection.

Equal-area

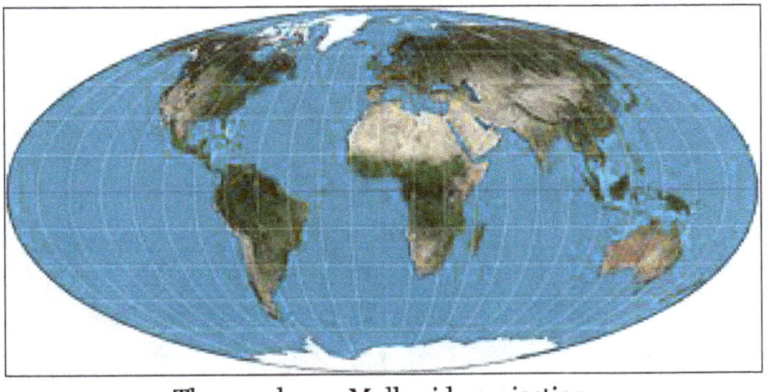

The equal-area Mollweide projection.

Equal-area maps preserve area measure, generally distorting shapes in order to do that. Equal-area maps are also called *equivalent* or *authalic*. These are some projections that preserve area:

- Albers conic,
- Bonne,
- Bottomley,
- Collignon,
- Cylindrical equal-area,
- Eckert II, IV and VI,
- Equal Earth,
- Gall orthographic (also known as Gall–Peters, or Peters, projection),
- Goode's homolosine,
- Hammer,
- Hobo–Dyer,
- Lambert azimuthal equal-area,
- Lambert cylindrical equal-area,

- Mollweide,
- Sinusoidal,
- Strebe,
- Snyder's equal-area polyhedral projection, used for geodesic grids,
- Tobler hyperelliptical,
- Werner.

Equidistant

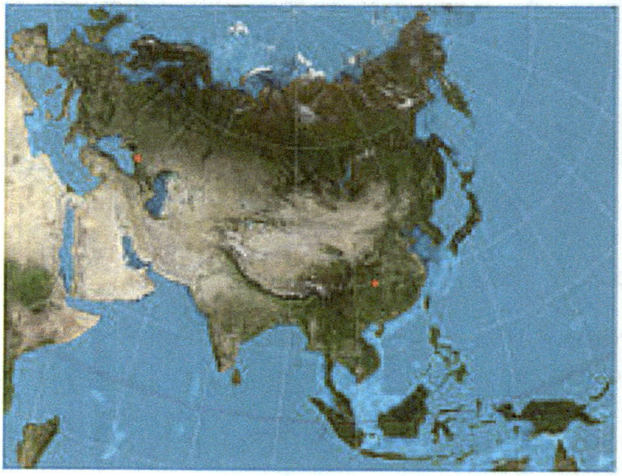

A two-point equidistant projection of Eurasia.

These are some projections that preserve distance from some standard point or line:

- Equirectangular—distances along meridians are conserved,
- Plate carrée—an Equirectangular projection centered at the equator,
- Azimuthal equidistant—distances along great circles radiating from centre are conserved,
- Equidistant conic,
- Sinusoidal—distances along parallels are conserved,
- Werner cordiform distances from the North Pole are correct as are the curved distance on parallels,
- Soldner,
- Two-point equidistant: two "control points" are arbitrarily chosen by the map maker. Distance from any point on the map to each control point is proportional to surface distance on the earth.

Gnomonic

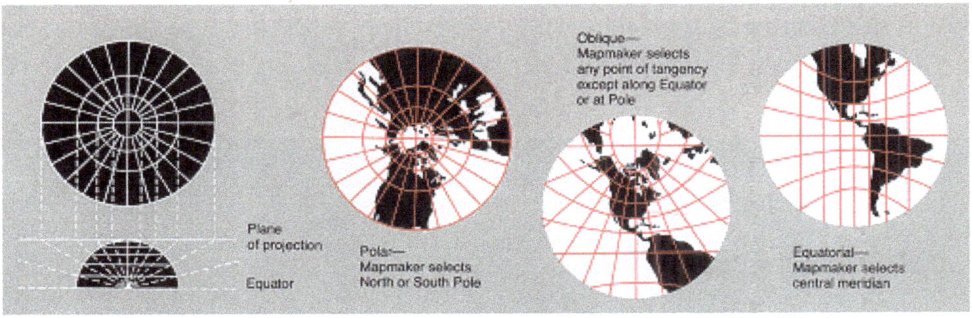

The Gnomonic projection is thought to be the oldest map projection.

Great circles are displayed as straight lines:

- Gnomonic projection.

Retroazimuthal

Direction to a fixed location B (the bearing at the starting location A of the shortest route) corresponds to the direction on the map from A to B:

- Littrow—the only conformal retroazimuthal projection,
- Hammer retroazimuthal—also preserves distance from the central point,
- Craig retroazimuthal *aka* Mecca or Qibla—also has vertical meridians.

Compromise Projections

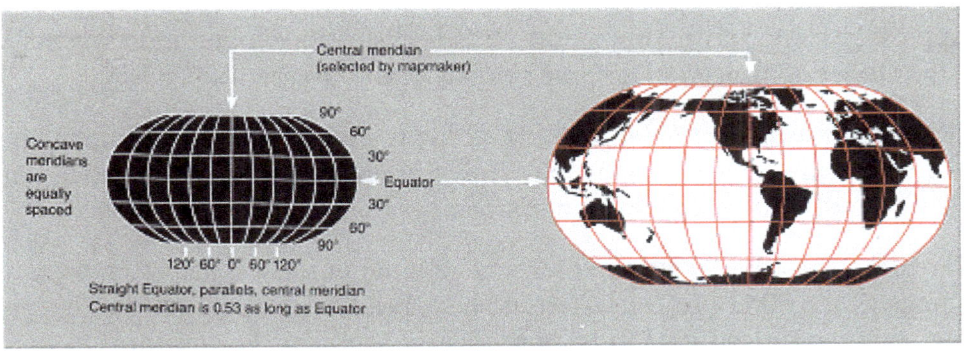

The Robinson projection was adopted by National Geographic magazine but abandoned by them for the Winkel tripel.

Compromise projections give up the idea of perfectly preserving metric properties, seeking instead to strike a balance between distortions, or to simply make things "look right". Most of these types of projections distort shape in the polar regions more than at the equator. These are some compromise projections.

- Robinson,
- Van der Grinten,
- Miller cylindrical,
- Winkel Tripel,
- Buckminster Fuller's Dymaxion,
- B. J. S. Cahill's Butterfly Map,
- Kavrayskiy VII projection,
- Wagner VI projection,
- Chamberlin trimetric,
- Oronce Finé's cordiform.

Best Projection

The mathematics of projection do not permit any particular map projection to be "best" for everything. Something will always be distorted. Thus, many projections exist to serve the many uses of maps and their vast range of scales.

Modern national mapping systems typically employ a transverse Mercator or close variant for large-scale maps in order to preserve conformality and low variation in scale over small areas. For smaller-scale maps, such as those spanning continents or the entire world, many projections are in common use according to their fitness for the purpose, such as Winkel tripel, Robinson and Mollweide. Reference maps of the world often appear on compromise projections. Due to distortions inherent in any map of the world, the choice of projection becomes largely one of aesthetics.

Thematic maps normally require an equal area projection so that phenomena per unit area are shown in correct proportion. However, representing area ratios correctly necessarily distorts shapes more than many maps that are not equal-area.

The Mercator projection, developed for navigational purposes, has often been used in world maps where other projections would have been more appropriate. This problem has long been recognized even outside professional circles. For example, a 1943 *New York Times* editorial states:

> "The time has come to discard (the Mercator) for something that represents the continents and directions less deceptively, Although its usage, has diminished, it is still highly popular as a wall map apparently in part because, as a rectangular map, it fills a rectangular wall space with more map, and clearly because its familiarity breeds more popularity".

A controversy in the 1980s over the Peters map motivated the American Cartographic Association (now Cartography and Geographic Information Society) to produce a series of booklets (including *Which Map Is Best*) designed to educate the public about map projections and distortion in maps. In 1989 and 1990, after some internal debate, seven North American geographic organizations adopted a resolution recommending against using any rectangular projection (including Mercator and Gall–Peters) for reference maps of the world.

CARTOGRAPHIC LABELING

Cartographic labeling is a form of typography and strongly deals with form, style, weight and size of type on a map. Essentially, labeling denotes the correct way to label features (points, arcs, or polygons).

Form

In type, form describes anything from lengths between letters to the case and color of the font. Form works well for both nominal (qualitative) and ordered (quantitative) data.

Italics

Italics describe the sloping of letters setting it apart from non-italicized words (or vice versa). Using italics on a map also slightly decreases the size of the font as it shapely squeezes it around features. When introduced, the idea was to condense the text by italicizing it, thus creating more text on the pages. The slope in the font was created to mimic the flow of cursive handwriting and thus, the angles of italic letters range anywhere from 11 to 30 degree and consequently, serifs are absent.

As a general rule on maps, the smaller the point size of a font, the more condensed and difficult it becomes to read. In an example of labeling a globe, ocean features are generally italicized to give an obvious discernment. In cartographic conventions, natural features are adequate in italics such as the aforementioned hydrographic features.

Case

Case is another way of emphasizing—whether it be uppercase, lowercase or a combination of the two (or even different size points within the same case). In general, uppercase fonts denote a higher emphasis, but according to Bringhurst (1996), an uppercase initial of a word has the seniority; but the lowercase letters have the control. In other words, the strong boldness of a larger letter draws the audience into its viewpoint. The lowercase letters contain the information needed to convey further. When viewing text on maps, it is still crucial to gain the audience's attention as a way of informing them of something other than the map(s). As for design, uppercase is much harder to read than mixed-use. In the globe example, mountain ranges should be in uppercase. When showing a larger scale, such as a region of the United States, it is useful to classify different case sizes. States should be in uppercase, with counties in small uppercase, and cities in lowercase.

Color

Color (value and hue) alterations also allow for a further emphasis on certain features. By changing the color of the font to correspond to the feature it is representing, the two become joined. If the cartographer were to label a river, the extra emphasis would be inherent if the font chosen was blue, to correspond with the blue feature (arc). On the contrary though, this is not always necessarily the case. If the cartographer chose a color of font for an ocean feature (polygon), blue would not be the obvious choice because it would appear to be washed out and thus, no emphasis. In this case, it is useful to label the feature with a more rich, bolder color (such as black font on blue polygon).

Spacing

The spacing of the letters on features also gives a more appealing map-visually speaking. By enlarging the increments between each letter of a word, the word in turn, becomes more pronounced. In the case for a long arc feature (river), to add more emphasis on the label, the letters would need to be extended or stretched. On the other hand, in some cases, the letters would have to be condensed (shortened increment gaps) to give a more proportional label for a feature.

Style

Serifs

The type style affects to overall look of the map and is adequately used to symbolize nominal (qualitative) data within the map. In general, style amounts to the use of serifs versus sans serifs. A serif is, by definition, a cross-line at the end of a stroke along a letter. On a map, the text that is chosen should be consistent. Generally, serif fonts are utilized to give a more regimented block body of text—similar to those used in traditional printing. Serifs are more widely used for historical information or a historical map.

Sans Serifs

The serif counterpart is sans serifs (meaning without serifs). Sans serif fonts are the more modern of the two fonts. But choosing one over the other requires that the audience will be able to read the text without strain. Generally, sans serifs are not for large bodies of text in print but instead, are ideal for the internet. On the same facet, sans serifs are optimal for a more-clean appearance in such places like a header, title, or legend. In map design, it's useful to also use sans serifs for natural features.

Weight

The type weight provides a substantial amount of emphasis of the cartographer's choosing. Weight is important because it involves the difference between bold and regular contrast. The degree of power that is increased with weight, must be proportional to the size of the letter. If not, a letter can be too intense and thus more difficult to read. Similarly, the spacing between the letters must be extended to provide adequate to read smoothly. Bold text creates direct attention to the eyes of the audience to pronounce certain information from cartographer.

Size

The type size of fonts stresses the importance and emphasis of the intended map. Size is expressed in points through the American point system with 1 point equaling 1/72" of vertical height. Furthermore, points also show the spacing between letters, words and lines. A larger size implies more importance or a greater relative quantity; smaller denotes less importance or less quantity. For design purposes, text using a size of less than 6 point is difficult to read. On the contraire, text that is larger than 26 point is too cumbersome for a standard-size paper format. For titles, a font larger than 10 point generally allows for a good working title. Also, it is important to use at least a 2-point difference between type sizes to allow the audience to see subtle changes.

Placement

With all of the type in order and adequately designed, the final step is the correct placement of labels. Placement describes each feature and its subsequent label(s). For area features, it is important to curve and extend the spaces to properly fill in the areas enough that the audience can discern different areas. As a cartographic convention, labels are usually as horizontal as possible with no upside-down labels. For line features, it is useful to allow the label to conform to the line pattern. Similar to a river (e.g. geographic features), the label should flow around the edges along the line being careful not to have the letters too extended. For point patterns, the minor patterns to follow include keeping labels on/in their respective features (e.g. coastal cities with labels on the land and not ocean). The major pattern for points is the placement along the point itself. The most widely accepted pattern is to start at the center and work outward towards the northeast quadrant from the point. Many studies have been researched to address the correct strategy for the placements. The point feature cartographic label placement (PFCLP) problem offers the solutions when point boxes overlap.

Many software features automatically choose label placements for the cartographer, but these are not always a fail-safe option. The use of good judgment and cartographic conventions are important to gain the best placement.

CHOROGRAPHY

Chorography is the art of describing or mapping a region or district, and by extension such a description or map. This term derives from the writings of the ancient geographer Pomponius Mela and Ptolemy, where it meant the geographical description of regions. However, its resonances of meaning have varied at different times. Richard Helgerson states that "chorography defines itself by opposition to chronicle. It is the genre devoted to place, and chronicle is the genre devoted to time". Darrell Rohl prefers a broad definition of "the representation of space or place".

In his text of the *Geographia* (2nd century CE), Ptolemy defined geography as the study of the entire world, but chorography as the study of its smaller parts—provinces, regions, cities, or ports. Its goal was "an impression of a part, as when one makes an image of just an ear or an eye"; and it dealt with "the qualities rather than the quantities of the things that it sets down". Ptolemy implied that it was a *graphic* technique, comprising the making of views (not simply maps), since he claimed that it required the skills of a draftsman or landscape artist, rather than the more technical skills of recording "proportional placements". Ptolemy's most recent English translators, however, render the term as "regional cartography".

Renaissance Revival

Ptolemy's text was rediscovered in the west at the beginning of the fifteenth century, and the term "chorography" was revived by humanist scholars. An early instance is a small-scale map of Britain in an early fifteenth-century manuscript, which is labelled a *tabula chorographica*. John Dee in 1570 regarded the practice as "an underling, and a twig of *Geographie*", by which the "plat" (plan or drawing) of a particular place would be exhibited to the eye.

The term also came to be used, however, for *written* descriptions of regions. These regions were extensively visited by the writer, who then combined local topographical description, summaries of the historical sources, and local knowledge and stories, into a text. The most influential example (at least in Britain) was probably William Camden's *Britannia*, which described itself on its title page as a *Chorographica descriptio*. William Harrison in 1587 similarly described his own "Description of Britaine" as an exercise in chorography, distinguishing it from the historical/chronological text of Holinshed's *Chronicles* (to which the "Description" formed an introductory section).

Peter Heylin in 1652 defined chorography as "the exact description of some Kingdom, Countrey, or particular Province of the same", and gave as examples Pausanias's *Description of Greece* (2nd century AD); Camden's *Britannia*; Lodovico Guicciardini's *Descrittione di tutti i Paesi Bassi* (on the Low Countries); and Leandro Alberti's *Descrizione d'Italia.*

William Camden.

Camden's *Britannia* was predominantly concerned with the history and antiquities of Britain, and, probably as a result, the term chorography in English came to be particularly associated with antiquarian texts. William Lambarde, John Stow, John Hooker, Michael Drayton, Tristram Risdon, John Aubrey and many others used it in this way, arising from a gentlemanly topophilia and a sense of service to one's county or city, until it was eventually often applied to the genre of county history. A late example was William Grey's *Chorographia*, a survey of the antiquities of the city of Newcastle upon Tyne. Even before Camden's work appeared, Andrew Melville in 1574 had referred to chorography and chronology as the "twa lights" [two lights] of history.

Example of Christopher Saxton's cartography.

However, the term also continued to be used for maps and map-making, particularly of sub-national or county areas. William Camden praised the county mapmakers Christopher Saxton and John Norden as "most skilfull Chorographers"; and Robert Plot in 1677 and Christopher Packe in 1743 both referred to their county maps as chorographies.

By the beginning of the eighteenth century the term had largely fallen out of use in all these contexts, being superseded for most purposes by either "topography" or "cartography". Samuel Johnson in his *Dictionary* made a distinction between geography, chorography and topography, arguing that geography dealt with large areas, topography with small areas, but chorography with intermediary areas, being "less in its object than geography, and greater than topography". In practice, however, the term is only rarely found in English by this date.

Modern Usages

Ferdinand von Richthofen.

In more technical geographical literature, the term had been abandoned as city views and city maps became more and more sophisticated and demanded a set of skills that required not only skilled draftsmanship but also some knowledge of scientific surveying. However, its use was revived for a second time in the late nineteenth century by the geographer Ferdinand von Richthofen. He regarded chorography as a specialization within geography, comprising the description through field observation of the particular traits of a given area.

The term is also now widely used by historians and literary scholars to refer to the early modern genre of topographical and antiquarian literature.

CARTOGRAPHIC GENERALIZATION

Cartographic generalization are the processes of selection and summarizing of the contents in drafting geographical maps. The purpose of such generalization is to preserve and distinguish on a map the main and typical outlines and the characteristic peculiarities of the features shown in accordance with the function, subject, and possible scales of the map. Map scale has the most obvious effect on cartographic generalization. For example, the representation of an area of 1 sq km on a map with a scale of 1:1000 will occupy an area of 1 sq m; with a scale of 1:10,000, 1 sq decimeter; with a scale of 1:100,000, 1 sq cm; and with a scale of 1:1,000,000, 1 sq mm. The depiction of a locality in all of these scales with identical detail and saturation is impossible. The exclusion of details and less important elements is inevitable as the scale becomes smaller. However, the effect of the scale is not only to limit the amount of space available on the map: on a small-scale map, which covers considerable area, details lose their significance and, if retained, would make it more difficult to perceive the main objects on the map. For example, an overall picture of the mountain systems of the Caucasus can be conveyed only on a small-scale and very generalized map, not on detailed topographical maps. Cartographic generalization is influenced by geographical conditions: the same features (or their peculiarities) are evaluated differently for different landscapes or according to the special nature of their relationship to other features—for example, wells are an important element in all topographical maps of desert and semidesert regions, but they are not indicated on similar maps showing areas with good water supply. Cartographic generalization is particularly affected by the function of the map. For example, in the case of a reference map, as much information as possible is provided, whereas on an educational map of the same scale the number of features shown will be reduced and limited to the requirements of the school program.

Cartographic generalization may appear in the following ways: (1) the selection of objects (the restriction of the contents of the map to objects that are essential), (2) the carefully considered simplification of contours (the planned outlining of objects, both linear and those that occupy an area, in which the peculiarities of the outlines typical of such objects are maintained and sometimes even emphasized—the sickle shape of oxbow lakes or the circular shape of lakes on out-wash plains), (3) the generalization of quantitative characteristics by reducing the number of divisions within which quantitative differences for specific features shown on the map are indicated (for example, in the case of a population scale for built-up areas, combining two divisions on the scale, such as "less than 500 inhabitants" and "from 500 to 2,000 inhabitants" into one division, "less than 2,000 inhabitants"), (4) the generalization of qualitative characteristics by simplifying the classifications for the features being shown (not subdividing forests according to type when showing vegetation on topographical maps (not subdividing forests according to type when showing vegetation on topographical maps), and (5) the replacement of individual features by general designations (indicating a population center by blocks and a geometrical sign instead of marking individual buildings).

The establishment of the principles governing cartographic generalization is an important scientific problem of cartography. An example is the establishment of rules of selection in a mathematical form, particularly in the form of quantitative indexes, the criteria that determine the conditions for indicating on the map objects of various categories (for example, the obligation to indicate all cities having 10,000 or more inhabitants). The selection indexes vary according to the map and the geographical region. The development of the mathematical bases of cartographic generalization has acquired considerable importance as a result of the introduction of automation into the processes of drafting and use of maps.

PHOTOGRAMMETRY

Photogrammetry is a technique that uses photographs for mapmaking and surveying. As early as 1851 the French inventor Aimé Laussedat perceived the possibilities of the application of the newly invented camera to mapping, but it was not until 50 years later that the technique was successfully employed. In the decade before World War I, terrestrial photogrammetry, as it came to be known later, was widely used; during the war the much more effective technique of aerial photogrammetry was introduced. Although aerial photogrammetry was used primarily for military purposes until the end of World War II, thereafter peacetime uses expanded enormously. Photography is today the principal method of making maps, especially of inaccessible areas, and is also heavily used in ecological studies and in forestry, among other uses.

From the air, large areas can be photographed quickly using special cameras, and blind areas, hidden from terrestrial cameras, are minimized. Each photograph is scaled, using marked and known ground reference points; thus, a mosaic can be constructed that may include thousands of photographs. Plotting machines and computers are used to overcome complications.

Instruments used in photogrammetry have become very sophisticated. Developments in the second half of the 20th century include satellite photography, very large scale photographs, automatic visual scanning, high-quality colour photographs, use of films sensitive to radiations beyond the visible spectrum, and numerical photogrammetry.

References

- About-geographic-coordinate-systems, map-projections, guide-books, 10.3, arcmap: arcgis.com, Retrieved 4 June, 2019
- Hughes-hallett, deborah; mccallum, william g.; gleason, andrew m. (2013). Calculus : single and multivariable (6 ed.). John wiley. Isbn 978-0470-88861-2
- Cartographic-considerations, fundamentals-mapping, education: gov.au, Retrieved 5 April, 2019
- T. Slocum, r. Mcmaster, f. Kessler, and h. Howard, thematic cartography and geographic visualization, 2nd edition, pearson, 2005, isbn 0-13-035123-7, p. 272
- Map-legend-in-printing-1078118: lifewire.com, Retrieved 3 February, 2019

- World meteorological organisation. “isallobar”. Eumetcal. Archived from the original on 16 april 2014. Retrieved 12 april 2014
- Scale-bars, layouts, help, pro-app: arcgis.com, Retrieved 26 August, 2019
- Dan reboussin (2005). Wind rose. Archived 2016-09-01 at the wayback machine university of florida. Retrieved on 2009-04-2
- Cartographic+generalization: thefreedictionary.com, Retrieved 13 March, 2019
- Cheng, y.; lorre, j. J. (2000). “equal area map projection for irregularly shaped objects”. Cartography and geographic information science. 27 (2): 91. Doi:10.1559/152304000783547957.
- Photogrammetry, science: britannica.com, Retrieved 1 January, 2019

3

Maps and its Types

Maps are the illustrations or the diagrammatical depictions of an area of land, sea or space, which portray physical features of cities, roads, etc. Some of the different types of maps are pictorial maps, nautical charts, world maps, reference maps, topographical maps, etc. This chapter has been carefully written to provide an easy understanding of these types of maps.

Map is the graphic representation, drawn to scale and usually on a flat surface, of features—for example, geographical, geological, or geopolitical—of an area of the Earth or of any other celestial body. Globes are maps represented on the surface of a sphere. Cartography is the art and science of making maps and charts.

In order to imply the elements of accurate relationships, and some formal method of projecting the spherical subject to a map plane, further qualifications might be applied to the definition. The tedious and somewhat abstract statements resulting from attempts to formulate precise definitions of maps and charts are more likely to confuse than to clarify. The words map, chart, and plat are used somewhat interchangeably. The connotations of use, however, are distinctive: charts for navigation purposes (nautical and aeronautical), plats (in a property-boundary sense) for land-line references and ownership, and maps for general reference.

Globe.

Cartography is allied with geography in its concern with the broader aspects of the Earth and its life. In early times cartographic efforts were more artistic than scientific and factual. As man explored and recorded his environment, the quality of his maps and charts improved.

Topographic maps are graphic representations of natural and man-made features of parts of the Earth's surface plotted to scale. They show the shape of land and record elevations above sea level, lakes, streams and other hydrographic features, and roads and other works of man. In short, they provide a complete inventory of the terrain and important information for all activities involving the use and development of the land. They provide the bases for specialized maps and data for compilation of generalized maps of smaller scale.

Nautical charts are maps of coastal and marine areas, providing information for navigation. They include depth curves or soundings or both; aids to navigation such as buoys, channel markers, and lights; islands, rocks, wrecks, reefs and other hazards; and significant features of the coastal areas, including promontories, church steeples, water towers, and other features helpful in determining positions from offshore.

The terms hydrography and hydrographer date from the mid-16th century; their focus has become restricted to studies of ocean depths and of the directions and intensities of oceanic currents; though at various times they embraced much of the sciences now called hydrology and oceanography. The British East India Company employed hydrographers in the 18th century, and the first hydrographer of the Royal Navy, Alexander Dalrymple, was appointed in 1795. A naval observatory and hydrographic office was established administratively in the United States Navy in 1854. In 1866 a hydrographic office was established by statute, and in 1962 it was renamed the U.S. Naval Oceanographic Office.

Interest in the charting of oceanic areas away from seacoasts developed in the second half of the 19th century, concurrently with the perfection of submarine cables. As knowledge of the configuration of the ocean basins increased, the attention of scientists was drawn to this field of study. A feature of marine science since the 1950s has been increasingly detailed bathymetric (water-depth measurement) surveys of selected portions of the seafloor. Together with collection of associated geophysical data and sampling of sediments, these studies assist in interpreting the geologic history of the ocean-covered portion of the Earth's crust.

Aeronautical charts provide essential data for the pilot and air navigator. They are, in effect, small-scale topographic maps on which current information on aids to navigation have been superimposed. To facilitate rapid recognition and orientation, principal features of the land that would be visible from an aircraft in flight are shown to the exclusion of less important details.

Modern Mapmaking Techniques

Compilation from Existing Materials

The preparation of derived maps—i.e., maps that are compiled from other maps or existing data—involves the search for, and evaluation of, all extant data pertaining to the subject area. Depending on the nature of the map to be compiled, thoroughgoing research includes boundary references, historical records, name derivations, and other materials. Selection of the most authentic items, on the frequent occasions when some ambiguities are detected, requires careful study and references to related materials. The sources finally selected may require some adjustment or compromise in order to fit properly with adjacent data. When it becomes evident that some sources are of questionable reliability, the cartographer explains this in the margin of his compilation. Sometimes this is placed in the body of the map where the doubtful features or delineations are located.

When selected materials have been assembled they are reduced to a common scale and copied on the compilation base, often in differentiating colours for the respective features. Reductions to a common scale are usually made by photography but may be made by projection and traced directly on the drawing. Minor adjustments may have to be made during compilation even though the source materials are of good quality. In particular, the need to make appropriate generalizations, omitting some details in smaller scale maps, requires much study and judgment.

Except for the new methods of preparing final colour-separation plates by scribing (described below), rather than by drafting or copperplate engraving, compilation processes have changed little over the years. Automatic-focusing projectors and better illumination have made the tracing of selected data at compilation scale easier. Better and more extensive facilities for photoreduction and copying, improved light tables, and a wider choice of drafting materials and instruments have served to facilitate compilation. The basic chores of research, selection of best data, and adjustment of these into the compilation, however, remain essentially the same.

The preparation of small-scale maps from large ones is sometimes simpler than the process just described, which pertains to compilation from a miscellany of differing sources. The relatively straightforward preparation of 1:62,500-scale maps from those of the 1:24,000-scale series, for example, may require little more than photoreduction and colour-separation drafting, or scribing. Even in this case some generalizations, as well as omission of a few of the least important details, are in order. To avoid the considerable expense involved in such scale conversions, straight photoreduction of colour-separation plates appears to be a promising procedure.

Larger reductions from one map series to another—1:62,500 to 1:250,000 for instance—are more of a problem, since the need for generalization is greater and the omission of many details is involved. The considerable differences in road and other symbol sizes also create displacement problems.

The component maps are reduced, and the negatives are cut and assembled into a mosaic on a clear sheet of plastic, the master negative of which provides guide copy for the several colour-separation plates required, which are then completed for reproduction. More often, however, it is necessary to make an intermediate compilation rather than burden the draftsman with too many adjustments to be made while following copy on the colour-separation plates. The intermediate scale for initial reduction of the component maps provides better legibility than direct photography to reproduction scale. This negative mosaic is copied on a metal-mounted drafting board. A compiler then inks the whole map, usually in three or more contrasting colours. He also draws roads and other symbols at the intermediate size, so that they will reduce to proper dimensions at reproduction scale, and makes the necessary displacement adjustments. Minor features and terrain details to be omitted on the new map are not inked in. The drawing is now ready for photoreduction to the final colour-separation plates, providing much better copy for the draftsman or engraver than direct reduction in one step would have produced.

Most smaller scale map series are prepared from large-scale maps as described above. In earlier days original reconnaissance surveys were made at small scales such as 1:192,000 for publication at 1:250,000. Ideally, the small-scale series of maps should be compiled progressively from those of larger scale and greater detail. Most countries, however, started their mapping programs with relatively small-scale reconnaissance surveys because of economic considerations. Later, affluence and technical competence permitted mapping at larger scales with better accuracy.

Geologic, soil, and other thematic maps usually have a topographic base from which woodland tints and road classification printings have been omitted. Such a map, therefore, has a topographic background printed in subdued colours on which the geologic or soil patterns are overprinted in prominent colours. Small-scale thematic maps showing weather patterns, vegetation types, and a large amount of economic and other information are of similar origin. Backgrounds are drawn from appropriate outline maps of provinces, countries, or regions of the world, while overlaying subject matters are compiled from specialized sources of information.

Generalization of Detail

The generalization of detail is a problem that frequently confronts the cartographer in original mapping and in reducing the scale of existing maps. There are two principal reasons for taking such liberties (or topographic license in the case of the original mapping). The primary purpose is to avoid overcrowding and the resulting poor legibility. In addition, the degree of generalization or detail should be as consistent as possible throughout the map. Generalizations in some parts and excessive detail in others confuse the user and make the map's reliability suspect. Effective generalization requires good judgment based on seasoned knowledge and experience.

In approaching such problems as the thousands of islets in the Stockholm archipelago or the thousands of small lakes in the Alaskan tundra areas, when the map scale will accommodate only a small number, the cartographer may decide to draw the features in groupings that reflect the patterns shown in the large-scale source maps or aerial photos. This is difficult and at best presents the nature of the respective areas rather than a literal portrayal. There is also the possibility that the source maps may already have been generalized by some omissions to accommodate to their own scales. Another device is to note, in appropriate text or marginal references, that many minor lakes or islets are omitted because of scale. Such areas may also be symbolized and explained. The "pattern" representation noted above is actually a form of symbolization.

Intricate coastlines are also extremely difficult to generalize consistently. Here again, the purpose is to omit minor details while retaining the main features and their distinguishing characteristics. These and many equally perplexing questions arise in preparing maps of very small scale from any source. The problems of equalization of detail are also present in such cases. The topographer of earlier days had the equalization problem between areas close at hand and those viewed distantly. In addition, the topographer had to deal with terrain on the far sides of obscuring features.

Photogrammetrists—that is, persons who compile original maps from aerial photos—have similar problems when, for example, one side of a ridge is seen in more detail than the opposite side. Indeed, in steep terrain, parts of the far sides of some mountains are not seen at all. Appropriate steps must be taken in such cases to avoid differing renditions on opposite sides of the mountain. This may be accomplished by adding, in field completion of the manuscript map, the segments not seen by the photogrammetrist; or additional aerial photography, patterned to cover the obscured sectors, may be requested.

Map Production from Original Surveys

The instrumentation, procedures, and standards involved in making original surveys have improved remarkably in recent years. Geodetic, topographic, hydrographic, and cadastral surveys have been facilitated by the application of electronics and computer sciences. At the same time, superior optics and more refined instruments, in general, have enhanced the precision of observations and accuracies of the end products.

The improved quality of surveys has increased the reliability of maps and charts based on them. In turn, the greater output of basic data has accelerated the production of maps and charts, while parallel improvements in processing steps have increased the volume and improved the final product. In a sense the production of maps from original surveys parallels the process steps after a compilation is made from derived sources. This phase is sometimes referred to as map finishing and involves editing, colour separation, and printing. In original surveys for topographic maps and nautical charts, however, the end products are provided for in all the process steps leading to the completed basic manuscript. The manuscript scale is, for example, selected to accommodate the plotting instruments involved as well as the final rendering for printing. In early years it was usual to choose a manuscript scale somewhat larger than that prescribed for publication. This was to allow for some generalization and line refinement in the final reduction. Thus, maps to be published at 1:62,500 scale were plotted in the field at 1:48,000 or thereabouts. With modern photogrammetric instruments, plotting is usually at reproduction scale.

Maps are not directly derived from geodetic surveys, and only land-line plats are produced from cadastral surveys. Accordingly, the primary original map and chart productions are those from topographic and hydrographic surveys. The surveys are somewhat similar as the nautical chart is, in effect, a topographic map of the coast with generalized offshore topography interpolated from depth soundings.

A variety of electronic devices are used to determine a survey ship's precise location while taking soundings, which are also made with electronic equipment. Both hydrographic and topographic surveys now employ aerial photography and precise plotting instruments to develop the base map. In order to simplify the description of modern mapmaking techniques, the process developed for topographic mapping will be described below, with comments where procedures for nautical charts differ significantly. Both processes start by expanding upon the basic control previously established from geodetic surveys.

Surveying, in which the facts are discovered and recorded, must precede mapping, in which the facts are presented in graphic form. Surveying involves (1) global positioning, in which the area to be mapped is located on the Earth's surface, usually by fixing a number of points in the area by astronomical observations or, after the techniques became available, by satellite or radar procedures; (2) establishing the framework, in which these points, and commonly many others connected by some combination of distance and angle measurements, are integrated into an accurately defined structure—like the steel framework of a modern building—on which the detail survey is based; and (3) making the detail survey, which establishes by less accurate (and therefore cheaper) methods the relative positions and shapes of the features being mapped. Constant reference to the framework prevents the errors in the detail survey from accumulating and growing unacceptably large.

Mapping also consists of three steps: (1) fair drawing, in which the accurate but not publishable surveyor's plot is redrawn by a skilled cartographer with uniform lines and lettering and, if a multicoloured map is being produced, is separated into several drawings, one for each colour; (2) reproduction, in which a negative is prepared from each of the fair-drawn originals and special colouring (to represent areas of vegetation, for example) is added; and (3) printing, in which a printing plate is made from each negative, the plates are mounted on a press, proofs (a few trial copies) are made to facilitate correction of errors and blemishes, and the final maps are produced.

Final Steps in Map Preparation

After all the features visible in the aerial photographs have been mapped, the manuscripts are contact printed on coated plastic sheets for review by the field engineer. He examines the whole map, adding such details as houses, trails, and fences that were not visible or were overlooked by the photogrammetrist. Political lines such as state, county, and township limits are located, as are geographic and other names in local use. Roads are classified, and woodland outlines are checked.

Contour accuracy is tested if the operator has noted areas that may be weak. The determination of names involves extensive local inquiry, as do political lines, and both may require research of records.

In remote areas it is more efficient to combine the above activities with supplemental control survey to avoid the extra field phase. Then the photos must be carefully examined and annotated for the compiler, while buildings must be encircled or pricked. Roads are classified and political lines located in the usual manner and noted on the photos or overlays.

Field corrections are applied to the original manuscripts. They must be scribed (engraved) on the originals so that guide copies can be prepared by contact printing for final colour-separation scribing. At this time all factual detail is carefully checked. Editing may proceed, to conserve time, while the colour-separation scribing is in progress. The editor reviews all names, boundaries, and related data, comparing them to information thereon that may be available from other sources. The editor's function is to see that the map conforms to standard conventions and is clear, legible, and free of errors.

Controversial names, or those found to be in confused or ambiguous spelling or usage, are documented and referred to an appropriate official body. The designation of type styles and sizes as well as placement of lettering is another function of the map editor.

Because modern topographic maps are printed in several colours, separate plates must be prepared for each. Some of the earliest maps were printed from woodcuts, usually in a single colour. Various hand processes were developed through the years, culminating in the fine rendering of copperplate engraving, which dominated the map production industry for many years. The process became obsolete, however, with increased production demands and the development of efficient printing presses. After World War II engraving on glass, and later on coated plastic sheets, was developed to a point that recovered the fineness of copper engraving. These methods of engraving have become firmly established in map production throughout the world.

Scribing

In the negative engraving or scribing process, guide copy is printed on several sheets of plastic coated with an opaque paint, usually yellow. The scriber follows copy on the respective plates by engraving through the coating. Because arc light can pass only through the engraving scratches, the completed engravings are, in effect, negatives from which the press plates are made. The finest lines (0.002 inch, or 0.05 millimetre, wide), such as intermediate contours, are engraved freehand. Heavier lines, such as index contours, engraved at 0.007 inch, may require a small tripod to assure that the scriber is perfectly vertical. Gravers for double-lined roads, others for buildings, and templates, or patterns, for a variety of symbols are used. Woodland and similar boundaries and shorelines are contact printed and etched on their respective coated sheets, and the areas of the woodland or water are then peeled off, leaving open windows for their respective features. If portions of scrub, orchard, or vineyard are contained in the "woodland" plate, negative sections for these are stripped into their respective locations. Press plates are then processed from the negatives.

A combined-colour proof is then made by successively printing the several completed negatives on a sensitized white plastic sheet that serves for the final checking and review of all aspects of the map. After all corrections have been made, the negatives are ready for the reproduction process.

Nearly all maps are now printed by rotary offset presses, using flexible aluminum-alloy printing plates. The system uses surface plates (very slightly raised or recessed) as opposed to the letterpress and intaglio processes, which involve greater image heights and depths respectively. In the printing sequence, ink goes from the plate to a rubber blanket to the paper. Thus, the printing plate is positive, or right-reading, as is the printed map. The negatives from which the printing plates are prepared are accordingly wrong-reading. This is the process for so-called surface plates. To retain fineness of line on very long runs (10,000 or more impressions), some map printers prefer "etched" plates, prepared from film positives. Both may be considered essentially surface plates, however, since the respective raise or recess is quite small.

Presses are of many varieties and makes. Huge multicolour types are used in large plants, printing several colours at a time. In effect, a multicolour press is several presses built into one. Each unit has three cylinders for plate, rubber blanket, and paper as well as rollers for water and ink. Presses with automatic feed may produce as many as 6,000 impressions per hour, while hand-fed types are limited to about 2,500 per hour.

Nautical Charts

Nautical charts are commonly large, 28 by 40 inches (70 centimetres by 1 metre) being an internationally accepted maximum size. In order that a navigator may work with them efficiently, charts must be kept with a minimum of folding in drawers in a large chart table in a compartment of the ship having ready access to the navigating bridge, known as the chart room or chart house. Such structures are not possible in small craft, which therefore require charts of a more convenient size. With the recognition that there are many more small boats in the world, particularly recreational craft, than there are ships and that they are navigated primarily by piloting rather than by celestial or electronic means, many hydrographic offices have given attention to the production of special chart series in a small format for yacht navigators.

A typical series is that produced by the U.S. Coast and Geodetic Survey with the designation SC (for small craft). Such charts are only 15 inches (38 centimetres) in the vertical dimension and thus need to be folded only in the vertical direction. Printed on both sides of the sheet, they are oriented along the most probable route rather than by parallels and meridians. Several are stapled together into a stiff cardboard folder for protection. Along with the ordinary chart information, they contain a year of tide tables and information on small-craft facilities in the area. New editions are produced annually.

Practical uses of charts impose some constraints on the selection of colour. Red, for example, would logically be chosen as the color in which to print warnings of navigational hazards. But navigators, who must work at night, prefer to retain the darkness adaptation of their eyes by viewing their charts under red light. Under such illumination, red, orange, and buff are invisible. Hence these colours have been superseded by magenta, purple, and gray.

Charts are working instruments, and, since ships often voyage far from where replacement charts are readily obtainable, hydrographic offices give attention to the quality of the paper on which charts are printed. A ship's reckoning is kept in pencil and erased after each voyage. Thus, printing stock that permits multiple erasures is chosen. In view of the environment where charts are used, another quality commonly sought is high wet strength.

Automation in Mapping

During the past few decades, there was much interest in the automation of mapping processes, and considerable progress was made in this area. Achievements in the fields of electronics, high-speed digital computers, and related technologies provided a favourable period for such progress. In Great Britain, development of a set of procedures utilizing automatic elements, known as the Oxford System, was begun.

Some success was also achieved in the difficult area of automatic plotting. Instruments now available can automatically scan a stereo model and generate approximate profiles from which contours may be interpolated. Some steps, however, must be closely monitored or else performed completely by the operator. Contouring interpolated from a profile scan is inferior to an operator's delineation. This contouring meets some less exacting requirements for elevation data, and refinements in the system are improving its precision. The need for human intervention when automatic devices get "lost" is not a decisive drawback, as one operator can monitor several machines. The reduction of tedious and repetitive steps for stereo-operators offers a significant advancement.

Coordinatographs with high repeat accuracies facilitate the automatic plotting of control points and projection intersections. Line work can also be drafted or scribed automatically by the same process, but the respective features must first be coded to provide the necessary input tapes. Automatic colour scanning and discrimination is operational but has not become widely used; it is still necessary for an operator to trace the various features on the manuscript to code them. Obviously, little is to be gained by automatic scribing until the input can be provided automatically. Coded line work can be displayed on a cathode-ray tube and corrected with a light pen, but it is much simpler to check and correct the manuscript or finished drawing. Systems of automatic type placement at present offer only marginal advantages over conventional methods. In short, automation has made substantial advances but has not become fully operational in a practical sense.

An aspect of automation that is developing rapidly concerns graphic data acquisition, storage, and retrieval. Data banks are being accumulated by specialized users of topographic information, often to produce thematic maps showing soil types, vegetation classifications, and a variety of other information. Such data banks are usually organized in two parts: one for line work, such as boundaries, and the other for descriptive information or classifications. Assuming that the necessary inputs have been made to the data bank, special plats can be generated speedily. Examples of such graphics include profiles showing elevations along a selected radio propagation path and cross sections for earthwork on roads and other construction.

ELEMENTS OF A MAP

Maps are the primary tools by which spatial relationships and geographic data are visualized. Maps therefore become important documents. There are several key elements that should be included each time a map is created in order to aid the viewer in understanding the communications of that map and to document the source of the geographic information used.

Parts of a Map

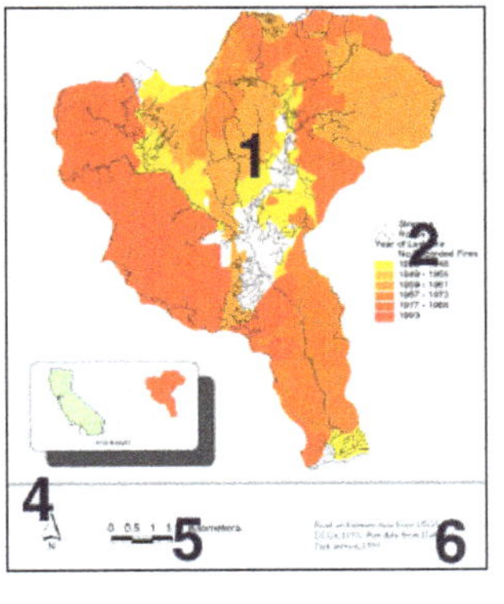

Numbered are descriptions of cartographic elements that are commonly found on a map layout. Some maps have all eight elements while other maps may only contain a few of them.

Data Frame

The data frame is the portion of the map that displays the data layers. This topic is the most important and central focus of the map document.

Legend

The legend serves as the decoder for the symbology in the data frame. Therefore, it is also commonly known as the key. In the legend below, the fire history schemata has been categorized with a graduating color scheme. The legend details which colors refer to which years. Without the legend, the color scheme on the map would make no sense to the viewer. The legend tells the viewer that the lighter the color, the longer the last recorded date of fire has been.

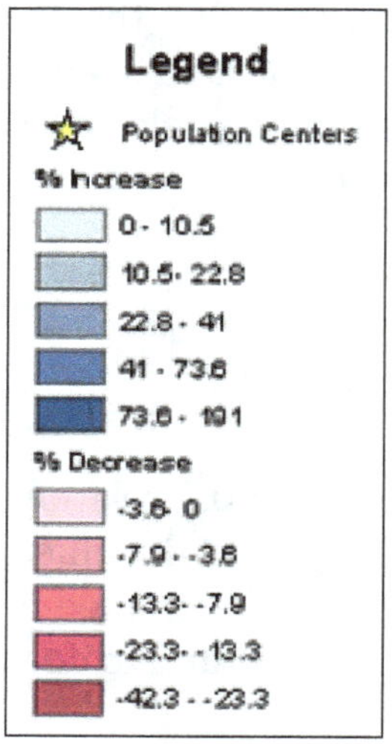

Title

The title is important because it instantly gives the viewer a succinct description of the subject matter of the map. The title "Fire History in Topanga, California" quickly tells the viewer the subject matter and location of the data.

North Arrow

The purpose of the north arrow is for orientation. This allows the viewer to determine the direction of the map as it relates to due north. Most maps tend to be oriented so that due north faces the top of the page. There are exceptions to this and having the north arrow allows the viewer to know which direction the data is oriented.

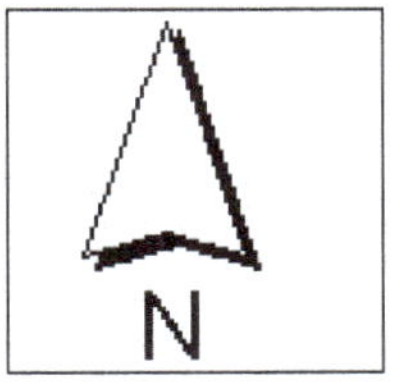

Scale

The scale explains the relationship of the data frame extent to the real world. The description is a ratio. This can be shown either as a unit to unit or as one measurement to another measurement. Therefore a scale showing a 1:10,000 scale means that everyone paper map unit represents 10,000 real world units. For example 1:10,000 in inches means that a measurement of one inch on the map equals 10,000 inches in real life. The second method of depicting scale is a comparison with different unit types. For example, 1″:100′ means that every inch measure on the paper map represents 100 feet in the real world. This ratio is the same as 1:1200 (1 foot = 12 inches). In addition to text representation as described above, the ratio can be shown graphically in the form of a scale bar. Maps that are not to scale tend have have a "N.T.S" notation which stands for "Not to scale."

Citation

The citation portion of a map constitutes the metadata of the map. This is the area where explanatory data about the data sources and currency, projection information and any caveats are placed. The citation tells the source and date of the data. Citations help the viewer determine the use of the map for their own purposes.

Linear Scale

A linear scale, also called a bar scale, scale bar, graphic scale, or graphical scale, is a means of visually showing the scaleof a map, nautical chart, engineering drawing, or architectural drawing.

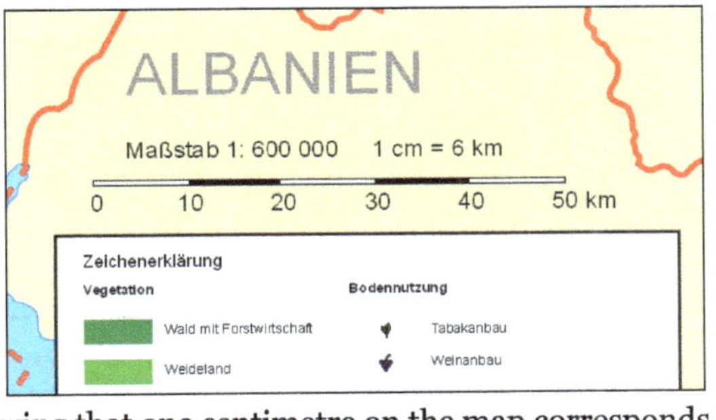

A linear scale showing that one centimetre on the map corresponds to six kilometres.

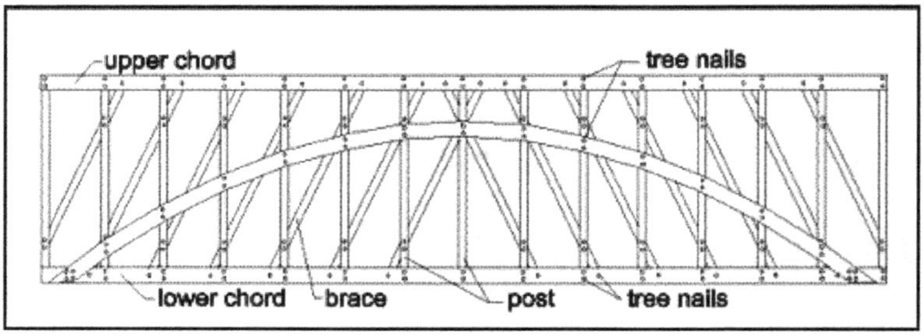

On large scale maps and charts, those covering a small area, and engineering and architectural drawings, the linear scale can be very simple, a line marked at intervals to show the distance on the earth or object which the distance on the scale represents. A person using the map can use a pair of dividers (or, less precisely, two fingers) to measure a distance by comparing it to the linear scale. The length of the line on the linear scale is equal to the distance represented on the earth multiplied by the map or chart's scale.

In most projections, scale varies with latitude, so on small scale maps, covering large areas and a wide range of latitudes, the linear scale must show the scale for the range of latitudes covered by the map.

Since most nautical charts are constructed using the Mercator projection whose scale varies substantially with latitude, linear scales are not used on charts with scales smaller than approximately 1/80,000. Mariners generally use the nautical mile, which, because a nautical mile is approximately equal to a minute of latitude, can be measured against the latitude scale at the sides of the chart.

While linear scales are used on architectural and engineering drawings, particularly those that are drawn after the subject has been built, many such drawings do not have a linear scale and are marked "Do Not Scale Drawing" in recognition of the fact that paper size changes with environmental changes and only dimensions that are specifically shown on the drawing can be used reliably in precise manufacturing.

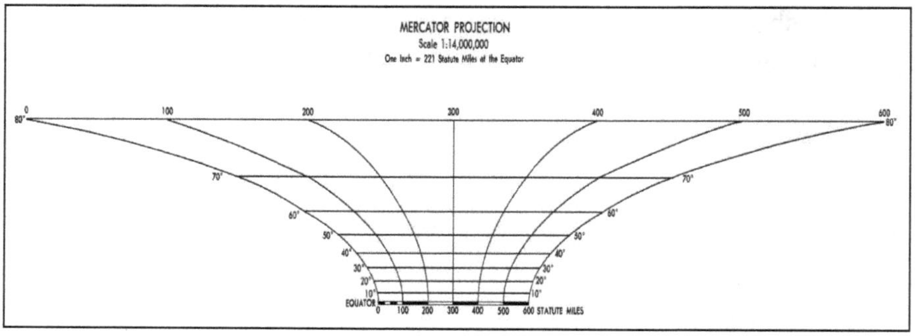

The scale from a large world map, showing, graphically, the change of scale with latitude. Each unit on the map at the equator represents the same distance on the earth as 5.9 units at 80° latitude.

PICTORIAL MAPS

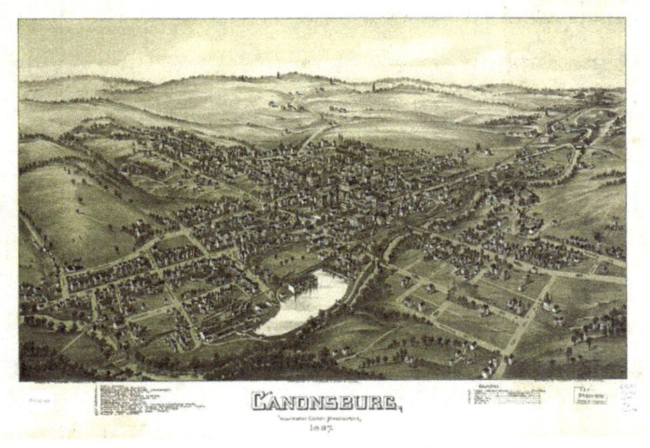

Pictorial map of Canonsburg.

Pictorial maps (also known as illustrated maps, panoramic maps, perspective maps, bird's-eye view maps, and geopictorial maps) depict a given territory with a more artistic rather than technical style. It is a type of map in contrast to road map, atlas, or topographic map. The cartography can be a sophisticated 3-D perspective landscape or a simple map graphic enlivened with illustrations of buildings, people and animals. They can feature all sorts of varied topics like historical events, legendary figures or local agricultural products and cover anything from an entire continent to a college campus. Drawn by specialized artists and illustrators, pictorial maps are a rich, centuries-old tradition and a diverse art form that ranges from cartoon maps on restaurant placemats to treasured art prints in museums.

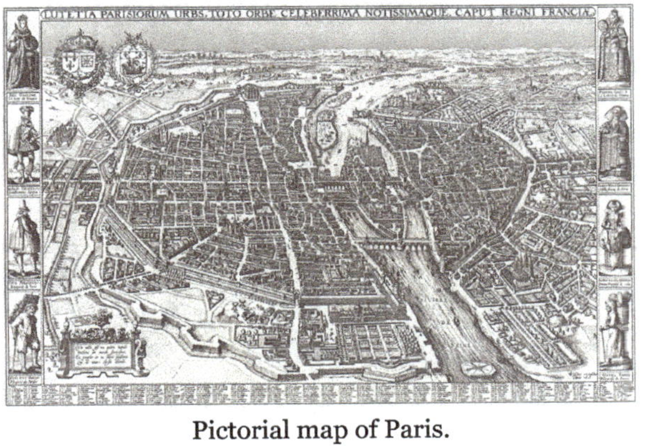

Pictorial map of Paris.

Pictorial maps usually show an area as if viewed from above at an oblique angle. They are not generally drawn to scale in order to show street patterns, individual buildings, and major landscape features in perspective. While regular maps focus on the accurate rendition of distances, pictorial maps enhance landmarks and often incorporate a complex interplay of different scales into one image in order to give the viewer a more familiar sense of recognition. With an emphasis on objects and style, these maps cover an artistic spectrum from childlike caricature to spectacular landscape graphic with the better ones being attractive, informative and highly accurate. Some require thousands of hours to produce.

Will Durant said that maps show us the face of history. Pictorial maps have been used to show the cuisine of a country, the industries of a city, the attractions of a tourist town, the history of a region or its holy shrines.

The history of pictorial maps overlaps much with the history of cartography in general and ancient artifacts suggest that pictorial mapping has been around since recorded history began.

In Medieval cartography, pictorial icons as well as religious and historical ideas usually overshadowed accurate geographic proportions. A classic example of this is the T and O map which represented the three known continents in the form of a cross with Jerusalem at its center. The more precise art of illustrating detailed bird's-eye-view urban landscapes flourished during the European Renaissance. As emerging trade centers such as Venice began to prosper, local rulers commissioned artists to develop pictorial overviews of their towns to help them organize trade fairs and direct the increasing flow of visiting merchants. When printing came around, pictorial maps evolved into some of the earliest forms of advertising as cities competed amongst themselves to attract larger shares of the known world's commerce.

Later, during the Age of Exploration, maps became progressively more accurate for navigation needs and were often sprinkled with sketches and drawings such as sailing ships showing the direction of trade winds, little trees and mounds to represent forests and mountains and of course, plenty of sea creatures and exotic natives much of them imaginary. As the need for geographical accuracy increased, these illustrations gradually slipped off the map and onto the borders and eventually disappeared altogether in the wake of modern scientific cartography.

Pictorial Map-makers up to Modern Times

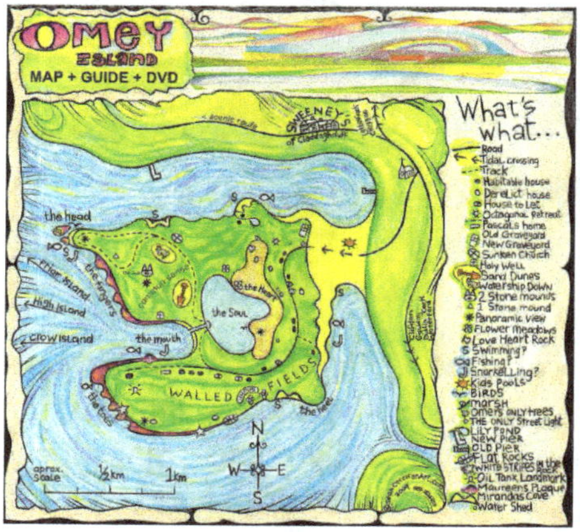

Colourful quirky map.

Ironically, despite all the changes that they record, very little has changed in the business of creating pictorial maps over the centuries. Showing off a given town, attracting visitors and stirring up local pride is what they have always been about. Most of these maps were and continue to be created by a handful of itinerant specialists who keep up the tradition. Many of them traveled from city to city enlisting the support of local merchants, industrialists and civic organizations whose endorsement would of course guarantee a prominent place for their properties on the map.

Edwin Whitefield for instance, one of the more prolific 19th-century American pictorial map artists, would require about 200 subscribers before he put pen to paper. Once he secured the profitability of the venture, Whitefield would be seen all over town furiously sketching every building. Then, choosing an imaginary aerial vantage point, he would integrate all his sketches into a complete and detailed drawing of the city. Then after that, say the chroniclers of the time, Whitefield would once again be seen furiously darting all over town to collect from all his sponsors. Says Jean-Louis Rheault, a contemporary pictorial map illustrator: "Pictorial maps - with their emphasis on what's important and eye-catching - make it easier to figure out what's where."

Tampa-Bay Aerial View Map.

Anthropomorphic Maps

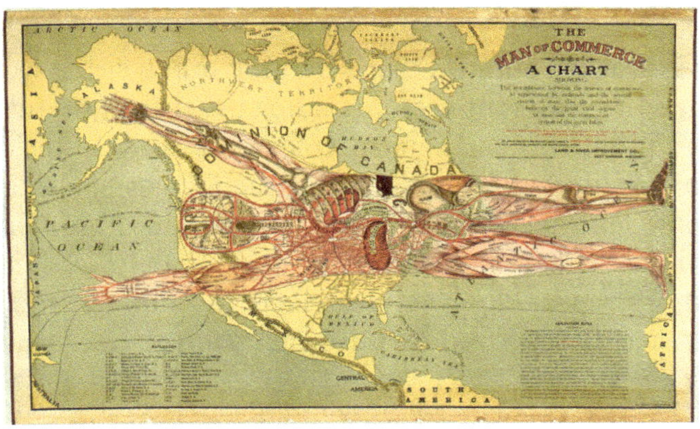

The Man of Commerce.

A type of pictoral maps are maps that use anthropomorphic images. Anthropomorphic maps date back to when Sebastian Münster used a queen to depict Europe in 1570. The map, *The Man of Commerce,* by Augustus F. McKay is the earliest anthropomorphic map known of in the United States, created in 1889.

GEOLOGIC MAPS

Geologists are rapidly incorporating GIS and information technology (IT) techniques into the production and dissemination of geologic maps.

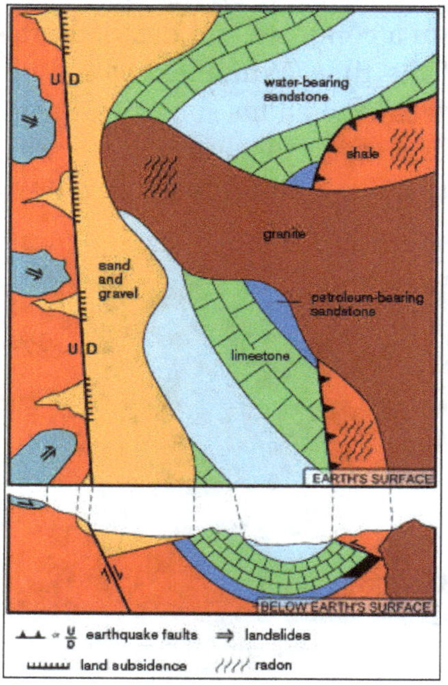

Graphic representation of typical information in a general purpose geologic map that can be used to identify geologic hazards, locate natural resources and facilitate land-use planning.

Geologic mapping is a highly interpretive, scientific process that can produce a range of map products for many different uses, including assessing ground-water quality and contamination risks; predicting earthquake, volcano, and landslide hazards; characterizing energy and mineral resources and their extraction costs; waste repository siting; land management and land-use planning; and general education. The value of geologic map information in public and private decision-making (such as for the siting of landfills and highways) has repeatedly been described anecdotally, and has been demonstrated in benefit-cost analyses to reduce uncertainty and, by extension, potential costs.

The geologic mapper strives to understand the composition and structure of geologic materials at the Earth's surface and at depth, and to depict observations and interpretations on maps using symbols and colors. Within the past 10 to 20 years, geographic information system (GIS) technology has begun to change aspects of geologic mapping by providing software tools that permit the geometry and characteristics of rock bodies and other geologic features (such as faults) to be electronically stored, displayed, queried, and analyzed in conjunction with a seemingly infinite variety of other data types.

For example, GIS can be used to spatially compare possible pollutant sources (such as oil wells) with nearby streams and geologic units that serve as ground-water supplies. In addition, GIS can be used to compare the position of a proposed road with the surrounding geology to identify areas of high excavation costs or unstable slopes. These comparisons have always been possible, but GIS greatly facilitates the analysis and, as a result, offers geologists the opportunity to provide information in map form that is easily interpreted and used by the nongeologist.

Geologic Mapping Field Methods

Geologists traditionally record in field notebooks their observations, including sketches, measurements (for example, the angle of tilted strata), and narratives. The validity of these observations remains; however, digital photographs now frequently supplement sketches, and instrumentation enhances measurement accuracy (for example, more precise locations are possible with a global positioning system instrument than with simple reference to position in relation to topographic and cultural features). Increasingly, field narratives are written and organized on a notebook computer or a personal digital assistant (PDA).

The development of field systems for recording and managing information has accelerated in the past 10 years. It is expected that the most useful systems will be widely adopted, thereby helping to standardize the techniques and formats for field observations. Increasingly, geologists not only are recording field observations but also are making field interpretations of the position of features such as geologic contacts and faults (that is, “drawing lines”) using rudimentary GIS software on PDAs or notebook computers.

Geologic Map Descriptors

When recording observations, geologists use descriptive terms and rock names that are in common use or unique to an area. These terms are then synthesized and rewritten into formal map unit descriptions that are published with the map. With the advent of GIS and the ease with which digital maps can be obtained and queried, geologists are recognizing the importance of a well-defined, standard terminology in order to help users, at a desktop computer or on the Internet, query simultaneously two maps made by different geologists. National and international efforts are now underway to define standard classifications for geologic information such as rock composition and texture. This standard language then can be used in the field or in the office to organize and interpret field observations. It is recognized that because of the many geologic terrains and geologic mapping agencies, and because of long historical usage of certain terms, multiple standard classifications will be necessary to accommodate regional variations in terminology. However, this system should function well, provided each classification is well defined and can be correlated with other classifications to ensure ease of translation from one format to another.

Geologic Map Databases

The main goal in many geological surveys no longer is to create a single geologic map but to create a database from which many types of geologic and engineering geology maps can be derived. This requires a database design or "data model" that is sufficiently robust to manage complex geologic concepts such as three dimensional (spatial) and temporal relations among map units, faults, and other features.

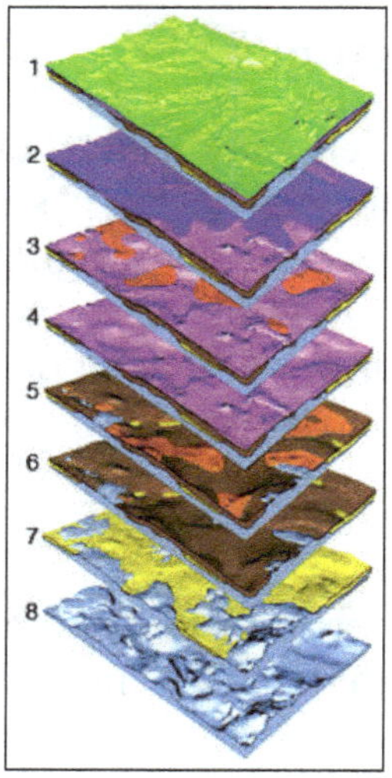

In figure, three-dimensional stack of glacial geologic layers in east-central Illinois. Layer 1 (top) is land surface; layer 8 (bottom) is the underlying bedrock. The light-colored unit in layer 7 is a sand and gravel aquifer filling a bedrock valley, and buried by low-permeability glacial till. After Soller, D.R., et al. Three-dimensional geologic maps of Quaternary sediments in East-Central Illinois, U.S. Geological Survey Geologic Investigation Series Map.

This is especially challenging because the software that manages these databases is not static but continues to evolve, thereby requiring an adaptable database design. Also, to permit the exchange of databases among agencies and users, either a common database design or a common interchange format is required. These design efforts are underway in North America and other parts of the world. As a consequence, geologists now are reevaluating how information is managed in the field, what kind of information is gathered and for what purpose, and the extent to which map information, once it has been created, can be updated to include new observations and interpretations.

Traditionally, once the geologist delivered a manuscript map to the cartographer, the job was finished. Map databases are meant to be maintained, and so a map can now be more accurately considered a progress report that can be updated. A critical aspect of the map database is the ability to manage the information and to preserve its integrity, for example, by migrating it to a new data format or structure or to a new standard terminology.

Cartography and Geologic Maps

GIS and graphic design software have radically changed the techniques by which map information is published. Traditional film (peel coat) technologies have been supplanted by digitally prepared negatives for offset printing of paper maps and by plot-on-demand approaches, which generate one map at a time upon request by users. Plot-on-demand is useful for producing customized, single-attribute maps (for example, showing only the shear strength of the geologic materials or the susceptibility of a groundwater supply to contamination). Digital cartographic techniques are evolving and becoming capable of producing sophisticated map layouts and products. It is anticipated that digital cartographers will be continually challenged to develop new techniques as software evolves and as geologists and users demand more complex and informative products.

Delivery of Geologic Maps and Databases

In addition to the conventional venues for obtaining maps (such as bookstores and sales offices), GIS and the Internet have made it possible to reach and educate new and potential users of geologic maps. Maps and databases now are available on the Internet in a variety of formats. Some formats (for example, Adobe Portable Document Format, or PDF) are designed for the visual display of a map and do not require specialized software, whereas other formats (for example, ArcInfo) are used by professionals with the appropriate software for the detailed analysis of a map database. To provide the public with access to such analyses without requiring them to purchase the software, numerous agencies are experimenting with software that permits users to view maps and to submit queries and view results within a Web browser. This technology is becoming increasingly widespread and should prove ever more useful as the following conditions are met:

- Increased availability of digital geology information.
- Standardized database structure, terminology, and interchange format.
- Increased Internet bandwidth.
- Advances in the software for serving map data.

New advances in mapping and preparation of map products made possible by advances in GIS and information technology have not altered the basic science of geology but offer new techniques for organizing, maintaining, and analyzing map data and potentially, for increasing its use by the public and by scientists.

NAUTICAL CHART

United States NOAA chart of part of Puerto Rico.

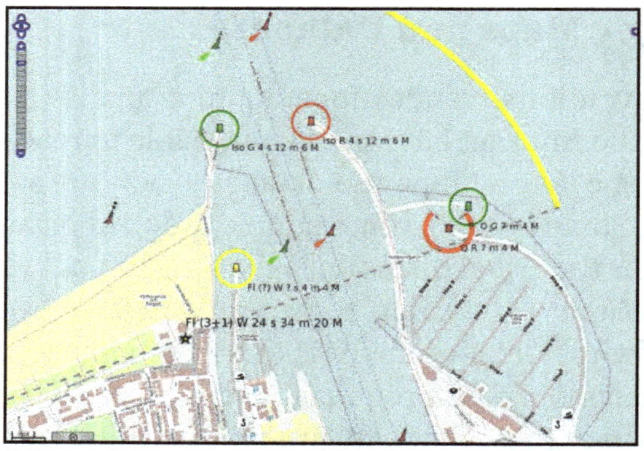

A nautical chart of the Warnemünde harbor shown on Open Sea Map.

A nautical chart is a graphic representation of a sea area and adjacent coastal regions. Depending on the scale of the chart, it may show depths of water and heights of land (topographic map), natural features of the seabed, details of the coastline, navigational hazards, locations of natural and human-made aids to navigation, information on tides and currents, local details of the Earth's magnetic field, and human-made structures such as harbours, buildings and bridges. Nautical charts are essential tools for marine navigation; many countries require vessels, especially commercial ships, to carry them. Nautical charting may take the form of charts printed on paper or computerized electronic navigational charts. Recent technologies have made available paper charts which are printed "on demand" with cartographic data that has been downloaded to the commercial printing company as recently as the night before printing. With each daily download, critical data such as Local Notices to Mariners are added to the on-demand chart files so that these charts are up to date at the time of printing.

Sources and Publication of Nautical Charts

Nautical charts are based on hydrographic surveys. As surveying is laborious and time-consuming, hydrographic data for many areas of sea may be dated and not always reliable. Depths are measured in a variety of ways. Historically the sounding line was used. In modern times, echo sounding is used for measuring the seabed in the open sea. When measuring the safe depth of water over an entire obstruction, such as a shipwreck, the minimum depth is checked by sweeping the area with a length of horizontal wire. This ensures that difficult to find projections, such as masts, do not present a danger to vessels navigating over the obstruction.

Nautical charts are issued by power of the national hydrographic offices in many countries. These charts are considered "official" in contrast to those made by commercial publishers. Many hydrographic offices provide regular, sometimes weekly, manual updates of their charts through their sales agents. Individual hydrographic offices produce national chart series and international chart series. Coordinated by the International Hydrographic Organization, the international chart series is a worldwide system of charts ("INT" chart series), which is being developed with the goal of unifying as many chart systems as possible.

There are also commercially published charts, some of which may carry additional information of particular interest, e.g. for yacht skippers.

Chart Correction

The nature of a waterway depicted by a chart may change, and artificial aids to navigation may be altered at short notice. Therefore, old or uncorrected charts should never be used for navigation. Every producer of nautical charts also provides a system to inform mariners of changes that affect the chart. In the United States, chart corrections and notifications of new editions are provided by various governmental agencies by way of Notice to Mariners, Local Notice to Mariners, Summary of Corrections, and Broadcast Notice to Mariners. In the U.S., NOAA also has a printing partner who prints the "POD" (print on demand) NOAA charts, and they contain the very latest corrections and notifications at the time of printing. To give notice to mariners, radio broadcasts provide advance notice of urgent corrections.

A good way to keep track of corrections is with a *Chart and Publication Correction Record Card* system. Using this system, the navigator does not immediately update every chart in the portfolio when a new *Notice to Mariners* arrives, instead creating a card for every chart and noting the correction on this card. When the time comes to use the chart, he pulls the chart and chart's card, and makes the indicated corrections on the chart. This system ensures that every chart is properly corrected prior to use. A prudent mariner should obtain a new chart if he has not kept track of corrections and his chart is more than several months old.

Various Digital Notices to Mariners systems are available on the market such as Digitrace, Voyager, or Chart Co, to correct British Admiralty charts as well as NOAA charts. These systems provide only vessel relevant corrections via e-mail or web downloads, reducing the time needed to sort out corrections for each chart. Tracings to assist corrections are provided at the same time.

The Canadian Coast Guard produces the Notice to Mariners publication which informs mariners of important navigational safety matters affecting Canadian Waters. This electronic publication is published on a monthly basis and can be downloaded from the Notices to Mariners (NOTMAR) Web site. The information in the Notice to Mariners is formatted to simplify the correction of paper charts and navigational publications.

Various and diverse methods exist for the correction of electronic navigational charts.

Limitations

In 1973 the cargo ship *MV Muirfield* (a merchant vessel named after Muirfield, Scotland) struck an unknown object in the Indian Ocean in waters charted at a depth of greater than 5,000 metres (16,404 ft), resulting in extensive damage to her keel. In 1983, HMAS *Moresby*, a Royal Australian Navy survey ship, surveyed the area where *Muirfield* was damaged, and charted in detail a previously unsuspected hazard to navigation, the Muirfield Seamount. The dramatic accidental discovery of the Muirfield Seamount is often cited as an example of limitations in the vertical geodetic datum accuracy of some offshore areas as represented on nautical charts, especially on small-scale charts.

A similar incident involving a passenger ship occurred in 1992 when the Cunard liner *Queen Elizabeth 2* struck a submerged rock off Block Island in the Atlantic Ocean. In November 1999, the semi-submersible, heavy-lift ship *Mighty Servant 2* capsized and sank after hitting an uncharted single underwater isolated pinnacle of granite off Indonesia. Five crew members died and *Mighty Servant 2* was declared a total loss. More recently, in 2005 the submarine USS *San Francisco* ran into an uncharted sea mount (sea mountain) about 560 kilometres (350 statute miles) south of Guam at a speed of 35 knots (40.3 mph; 64.8 km/h), sustaining serious damage and killing one seaman. In September 2006 the jack-up barge *Octopus* ran aground on an uncharted sea mount within the Orkney Islands (United Kingdom) while being towed by the tug *Harold*. £1M worth of damage was caused to the barge and delayed work on the installation of a tidal energy generator prototype. As stated in the Mariners Handbook and subsequent accident report: "No chart is infallible. Every chart is liable to be incomplete".

Map Projection, Positions and Bearings

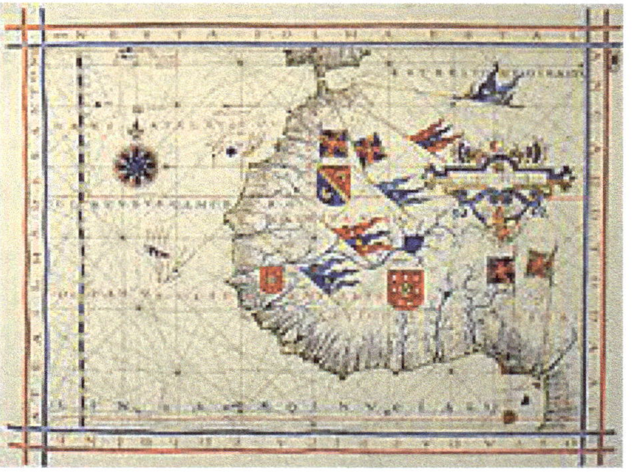

A pre-Mercator nautical chart of 1571, from Portuguese cartographer Fernão Vaz Dourado.

It belongs to the so-called *plane chart* model, where observed latitudes and magnetic directions are plotted directly into the plane, with a constant scale, as if the Earth's surface were a flat plane.

The Mercator projection is used on the vast majority of nautical charts. Since the Mercator projection is conformal, that is, bearings in the chart are identical to the corresponding angles in nature, courses plotted on the chart may be used directly as the course-to-steer at the helm.

The gnomonic projection is used for charts intended for plotting of great circle routes. NOAA uses the polyconic projection for some of its charts of the Great Lakes, at both large and small scales.

Positions of places shown on the chart can be measured from the longitude and latitude scales on the borders of the chart, relative to a geodetic datum such as WGS 84.

A bearing is the angle between the line joining the two points of interest and the line from one of the points to the north, such as a ship's course or a compass reading to a landmark. On nautical charts, the top of the chart is always true north, rather than magnetic north, towards which a compass points. Most charts include a compass rose depicting the variation between magnetic and true north.

However, the use of the Mercator projection has drawbacks. This projection shows the lines of longitude as parallel. On the real globe, the lines of longitude converge as they approach the north or south pole. This means that east-west distances are exaggerated at high latitudes. To keep the projection conformal, the projection increases the displayed distance between lines of latitude (north-south distances) in proportion; thus a square is shown as a square everywhere on the chart, but a square on the Arctic Circle appears much bigger than a square of the same size at the equator. In practical use, this is less of a problem than it sounds. One minute of latitude is, for practical purposes, a nautical mile. Distances in nautical miles can therefore be measured on the latitude gradations printed on the side of the chart.

Electronic and Paper Charts

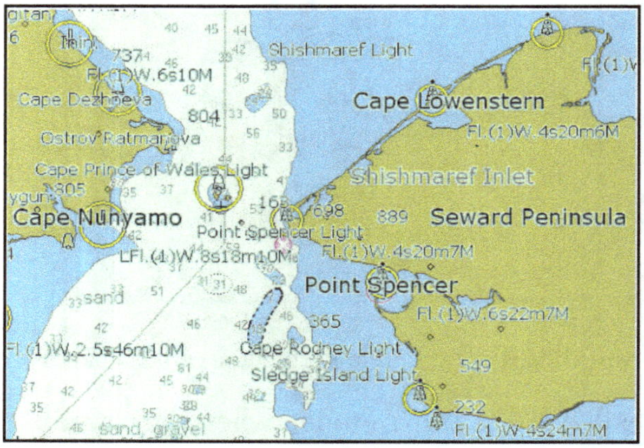

Portion of an electronic chart.

Conventional nautical charts are printed on large sheets of paper at a variety of scales. Mariners will generally carry many charts to provide sufficient detail for the areas they might need to visit. Electronic navigational charts, which use computer software and electronic databases to provide navigation information, can augment or in some cases replace paper charts, though many mariners carry paper charts as a backup in case the electronic charting system fails.

Labeling Nautical Charts

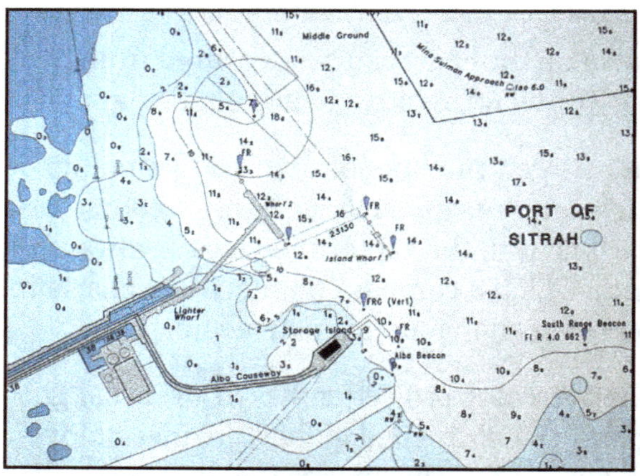

Automatically labeled nautical chart.

Nautical charts must be labeled with navigational and depth information. There are a few commercial software packages that do automatic label placement for any kind of map or chart.

Details on a Nautical Chart

Many countries' hydrographic agencies publish a "Chart 1", which explains all of the symbols, terms and abbreviations used on charts that they produce for both domestic and international use. Each country starts with the base symbology specified in IHO standard INT 1, and is then permitted to add its own supplemental symbologies to its domestic charts, which are also explained in its version of Chart 1. Ships are typically required to carry copies of Chart 1 with their paper charts.

Pilotage Information

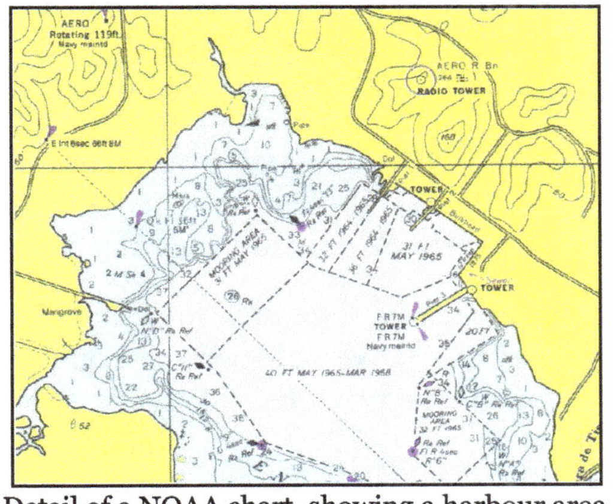

Detail of a NOAA chart, showing a harbour area.

The chart uses symbols to provide pilotage information about the nature and position of features useful to navigators, such as sea bed information, sea mark, and landmarks. Some symbols describe the sea bed with information such as its depth, materials as well as possible hazards such as shipwrecks. Other symbols show the position and characteristics of buoys, lights, lighthouses, coastal and land features and structures that are useful for position fixing. The abbreviation "ED" is commonly used to label geographic locations whose existence is doubtful.

Colours distinguish between man-made features, dry land, sea bed that dries with the tide, and seabed that is permanently underwater and indicate water depth.

Depths and Heights

Depths which have been measured are indicated by the numbers shown on the chart. Depths on charts published in most parts of the world use metres. Older charts, as well as those published by the United States government, may use feet or fathoms. Depth contour lines show the shape of underwater relief. Coloured areas of the sea emphasise shallow water and dangerous underwater obstructions. Depths are measured from the chart datum, which is related to the local sea level. The chart datum varies according to the standard used by each national Hydrographic Office. In general, the move is towards using lowest astronomical tide (LAT), the lowest tide predicted in the full tidal cycle, but in non-tidal areas and some tidal areas Mean Sea Level (MSL) is used.

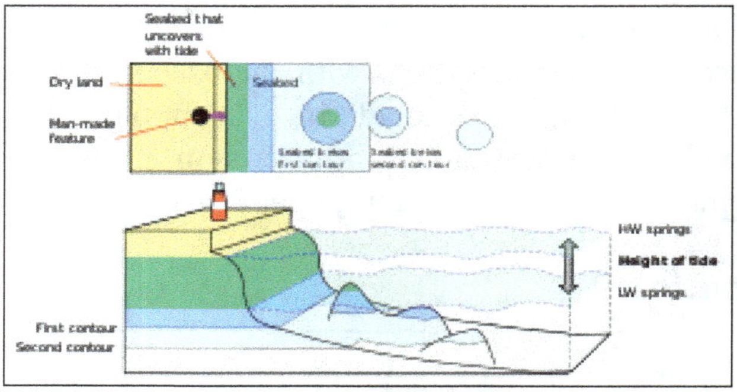

Use of colour in British Admiralty charts.

Heights, e.g. a lighthouse, are generally given relative to mean high water spring (MHWS). Vertical clearances, e.g. below a bridge or cable, are given relative to highest astronomical tide (HAT). The chart will indicate what datum is in use.

The use of HAT for heights and LAT for depths, means that the mariner can quickly look at the chart to ensure that they have sufficient clearance to pass any obstruction, though they may have to calculate height of tide to ensure their safety.

Tidal Information

Tidal races and other strong currents have special chart symbols. Tidal flow information may be shown on charts using tidal diamonds, indicating the speed and bearing of the tidal flow during each hour of the tidal cycle.

DYMAXION MAP

The Dymaxion map or Fuller map is a projection of a world map onto the surface of an icosahedron, which can be unfolded and flattened to two dimensions. The flat map is heavily interrupted in order to preserve shapes and sizes.

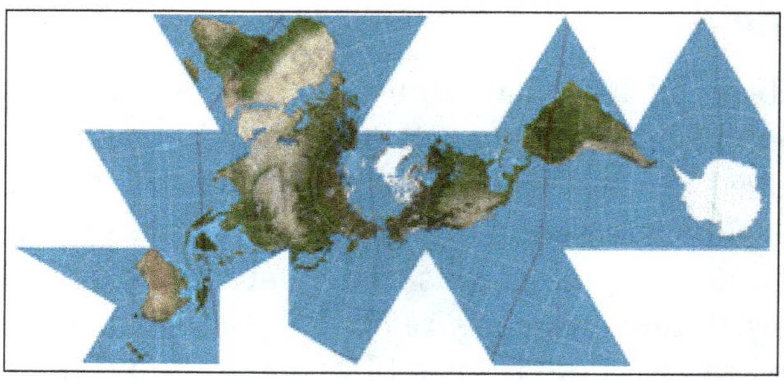

The world on a Dymaxion projection, with 15° graticule.

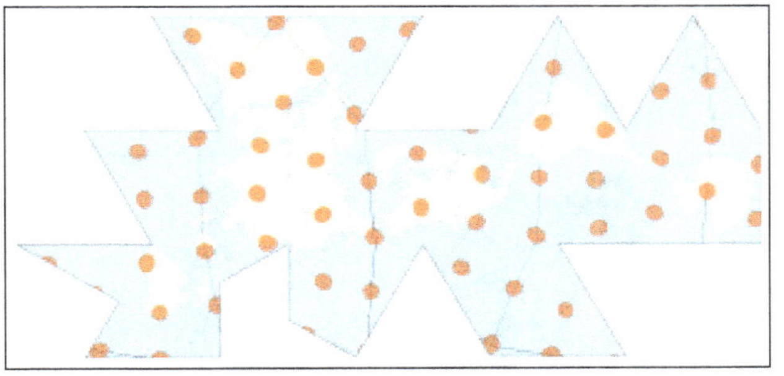

The Dymaxion projection with Tissot's indicatrix of deformation.

The projection was invented by Buckminster Fuller. The March 1, 1943 edition of *Life* magazine included a photographic essay titled "Life Presents R. Buckminster Fuller's Dymaxion World". Fuller applied for a patent in the United States in February 1944, the patent application showing a projection onto a cuboctahedron. The patent was issued in January 1946.

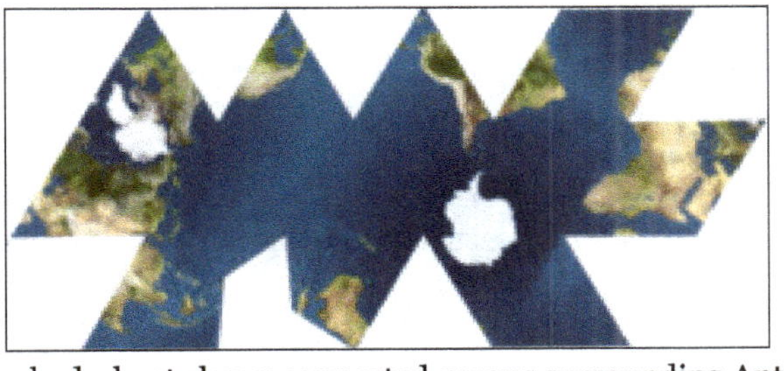

This icosahedral net shows connected oceans surrounding Antarctica.

The 1954 version published by Fuller, made with co-cartographer Shoji Sadao, the Airocean World Map, used a modified but mostly regular icosahedron as the base for the projection, which is the version most commonly referred to today. This version depicts the Earth's continents as "one island", or nearly contiguous land masses.

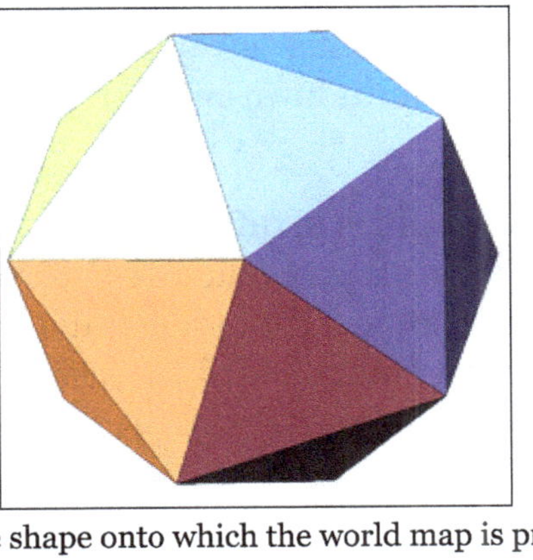

An icosahedron: This is the shape onto which the world map is projected before unfolding.

The Dymaxion projection is intended only for representations of the entire globe. It is not a gnomonic projection, whereby global data expands from the center point of a tangent facet outward to the edges. Instead, each triangle edge of the Dymaxion map matches the scale of a partial great circle on a corresponding globe, and other points within each facet shrink toward its middle, rather than enlarging to the peripheries.

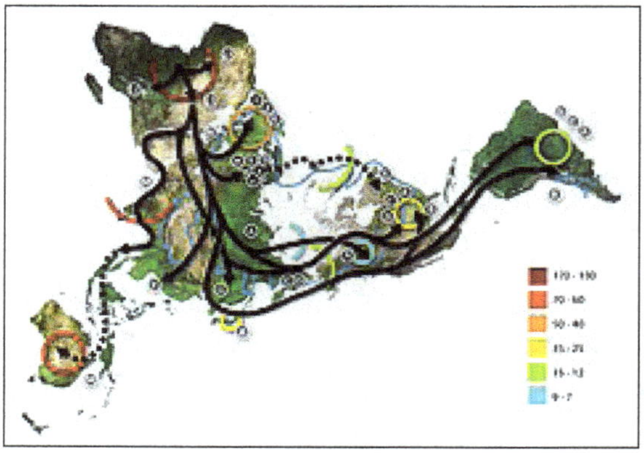

Example of use illustrating early human migrations according to mitochondrialpopulation genetics (numbers are millenniabefore present).

The name *Dymaxion* was applied by Fuller to several of his inventions.

Properties

Fuller claimed that his map had several advantages over other projections for world maps.

It has less distortion of relative size of areas, most notably when compared to the Mercator projection; and less distortion of shapes of areas, notably when compared to the Gall–Peters projection. Other compromise projection sattempt a similar trade-off.

More unusually, the Dymaxion map does not have any "right way up". Fuller argued that in the universe there is no "up" and "down", or "north" and "south": only "in" and "out". Gravitational forces of the stars and planets created "in", meaning "towards the gravitational center", and "out", meaning "away from the gravitational center". He attributed the north-up-superior/south-down-inferior presentation of most other world maps to cultural bias.

Fuller intended the map to be unfolded in different ways to emphasize different aspects of the world. Peeling the triangular faces of the icosahedron apart in one way results in an icosahedral net that shows an almost contiguous land mass comprising all of Earth's continents – not groups of continents divided by oceans. Peeling the solid apart in a different way presents a view of the world dominated by connected oceans surrounded by land.

Showing the continents as "one island earth" also helped Fuller explain, in his book *Critical Path*, the journeys of early seafaring people, who were in effect using prevailing winds to circumnavigate this world island.

However, the Dymaxion map can also prove difficult to use. It is, for example, confusing to describe the four cardinal directions and locate geographic coordinates. The awkward shape of the map may be counterintuitive to most people trying to use it. For example, tracing a path from India to Chile may be confusing. Depending on how the map is projected, land masses and oceans are often divided into several pieces.

Impact

A 1967 Jasper Johns painting, *Map (Based on Buckminster Fuller's Dymaxion Air-ocean World)*, depicting a Dymaxion map, hangs in the permanent collection of the Museum Ludwig in Cologne.

The World Game, a collaborative simulation game in which players attempt to solve world problems, is played on a 70-by-35-foot Dymaxion map.

In 2013, to commemorate the 70th anniversary of the publication of the Dymaxion map in *Life* magazine, the Buckminster Fuller Institute announced the "Dymax Redux", a competition for graphic designers and visual artists to re-imagine the Dymaxion map. The competition received over 300 entries from 42 countries.

WORLD MAP

The world, Abraham Ortelius's *Typus Orbis Terrarum*.

A world map is a map of most or all of the surface of Earth. World maps form a distinctive category of maps due to the problem of projection. Maps by necessity distort the presentation of the earth's surface. These distortions reach extremes in a world map. The many ways of projecting the earth reflect diverse technical and aesthetic goals for world maps.

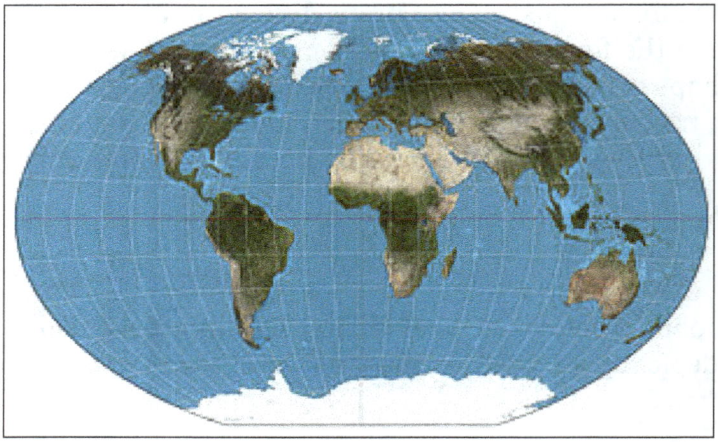

A world map on the Winkel tripel projection, a low-error map projection.

World maps are also distinct for the global knowledge required to construct them. A meaningful map of the world could not be constructed before the European Renaissance because less than half of the earth's coastlines, let alone its interior regions, were known to any culture. New knowledge of the Earth's surface has been accumulating ever since and continues to this day.

Maps of the world generally focus either on political features or on physical features. Political maps emphasize territorial boundaries and human settlement. Physical maps show geographic features such as mountains, soil typeor land use. Geological maps show not only the surface, but characteristics of the underlying rock, fault lines, and subsurface structures. Choropleth maps use color hue and intensity to contrast differences between regions, such as demographic or economic statistics.

Map Projections

A map is made using a map projection, which is any method of representing a globe on a plane. All projections distort distances and directions, and each projection distributes those distortions differently. Perhaps the most well known projection is the Mercator Projection, originally designed as a nautical chart.

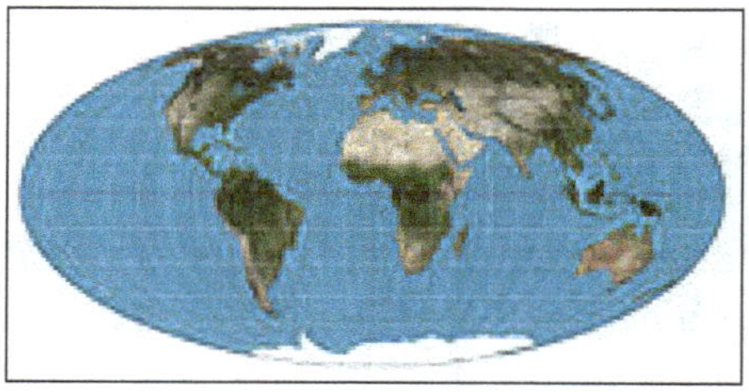

Mollweide projection.

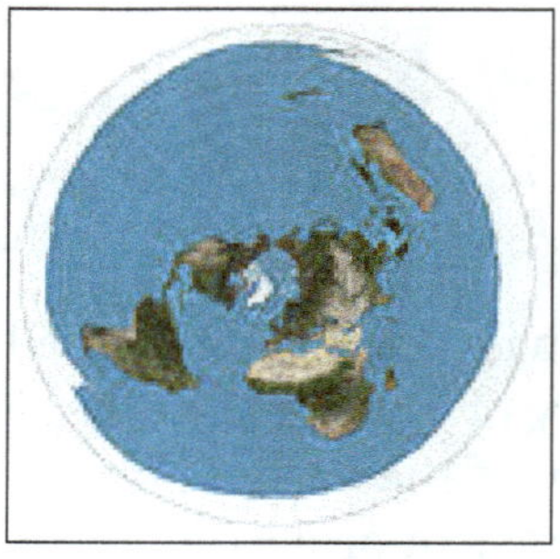

Polar azimuthal equidistant projection.

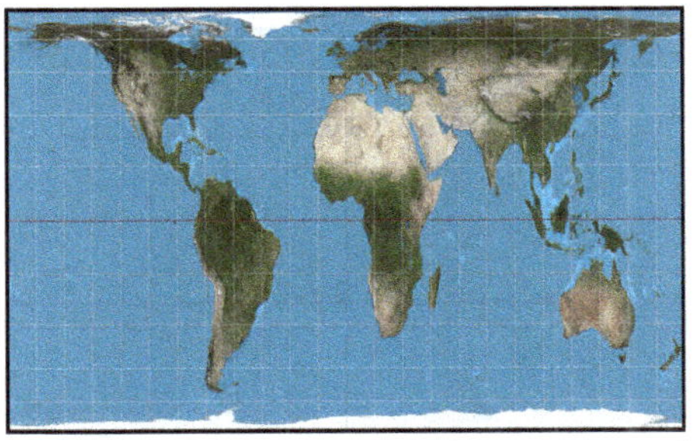

Gall–Peters projection, an equal area map projection.

Thematic Maps

A thematic map shows geographic information about one or a few focused subjects. These maps "can portray physical, social, political, cultural, economic, sociological, agricultural, or any other aspects of a city, state, region, nation, or continent".

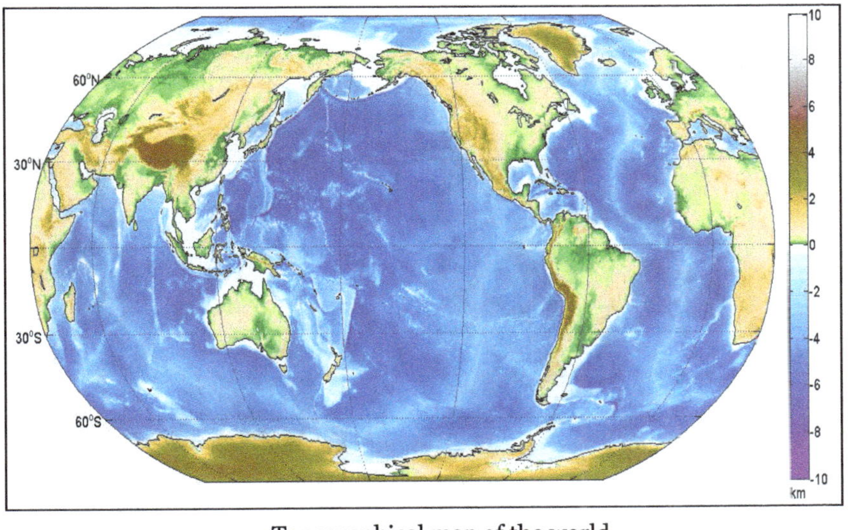

Topographical map of the world.

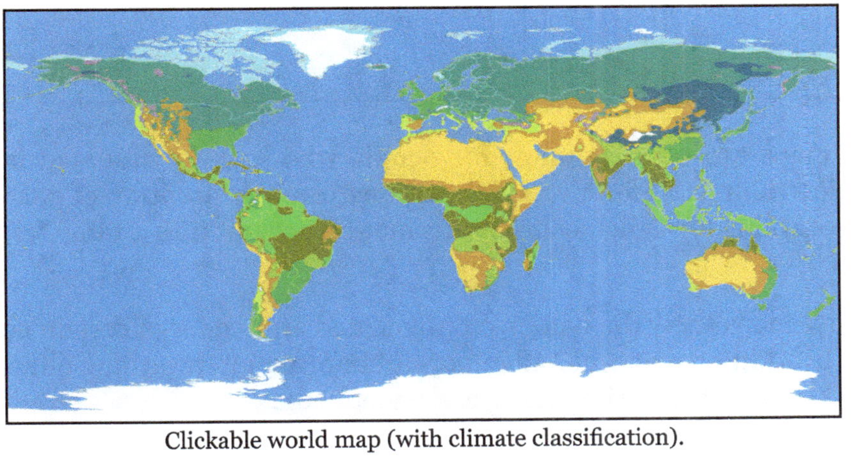

Clickable world map (with climate classification).

Historical Maps

World map according to Posidonius, drawn in 1628.

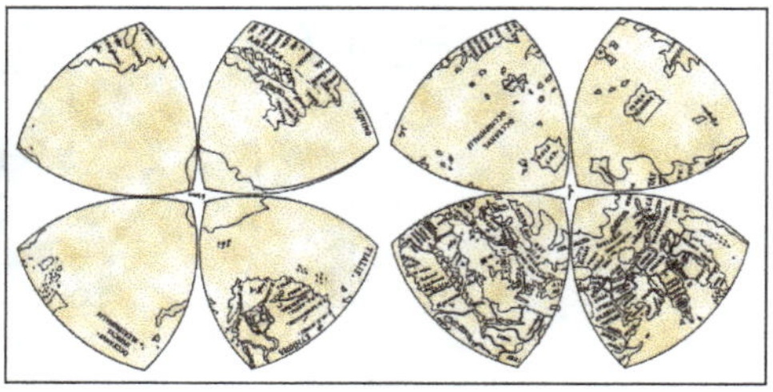

World map in Octant projection. From Leonardo da Vinci's Windsor papers.

Early world maps cover depictions of the world from the Iron Age to the Age of Discovery and the emergence of modern geography during the early modern period. Old maps provide much information about what was known in times past, as well as the philosophy and cultural basis of the map, which were often much different from modern cartography. Maps are one means by which scientists distribute their ideas and pass them on to future generations.

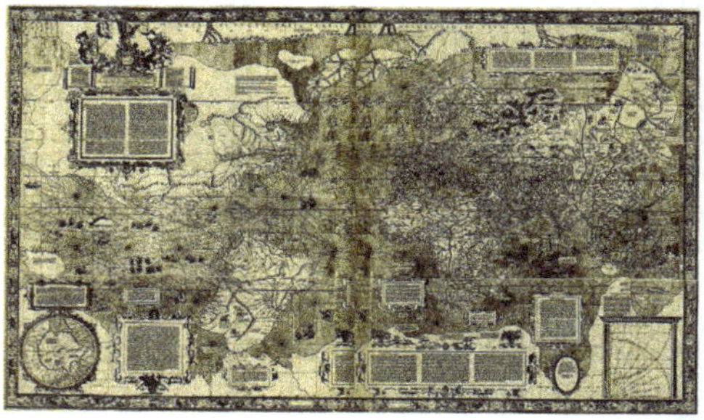

World map by Gerardus Mercator, first map in the well known Mercator projection.

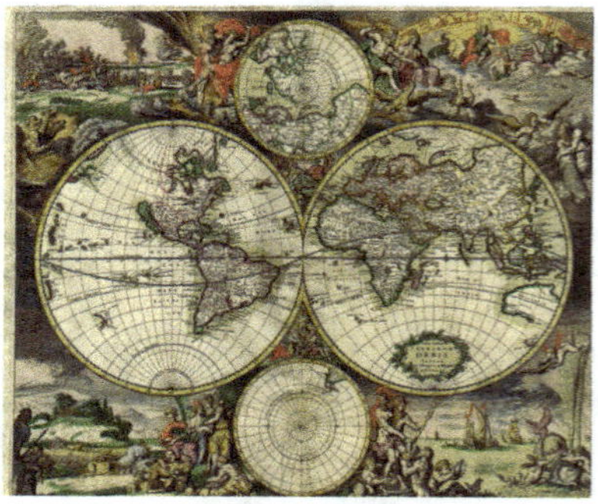

A historical map of the world by Gerard van Schagen.

REFERENCE MAPS

Reference maps are popular maps for people who look for updated information and statistics/figures. Reference maps demonstrate the situation of particular geographical locations for which census information are put into a table (charted) and distributed.

The reference maps show the names, frontiers, and area codes of certain geographical territories. In addition, you will find important information regarding physical aspects like railways, roads, rivers, shorelines, and water bodies.

Types of reference maps the reference maps can be categorized into the following types:

- Maps and information on a particular country,
- Geography maps,
- History maps.

The maps and information on a specific country can include information on the educational institutes, business centers, stock exchanges, oil refineries, languages, politics, assembly and parliamentary constituencies.

The geography maps can be further subcategorized into elevation maps, maps of neighboring countries, latitude and longitude maps, topographic map, seismic zone map, climatic region map, natural hazard map, soil map, river map, vegetation map, crop map, and agricultural maps.

The demography maps can be additionally categorized into clickable population maps, population density map, literacy map, and clickable sex ratio map.

The history maps include history map in Flash and maps of different periods of history. Continental maps and time zone maps can also be included in the category of reference maps. The different formats of reference map include Portable Document Format (PDF) and Graphics Interchange Format (GIF).

TOPOGRAPHIC MAP

A topographic map is a detailed and accurate two-dimensional representation of natural and human-made features on the Earth's surface. These maps are used for a number of applications, from camping, hunting, fishing, and hiking to urban planning, resource management, and surveying. The most distinctive characteristic of a topographic map is that the three-dimensional shape of the Earth's surface is modeled by the use of contour lines. Contours are imaginary lines that connect locations of similar elevation. Contours make it possible to represent the height of mountains and steepness of slopes on a two-dimensional map surface. Topographic maps also use a variety of symbols to describe both natural and human made features such as roads, buildings, quarries, lakes, streams, and vegetation.

Topographic Map Symbols

Topographic maps use symbols to represent natural and human constructed features found in the environment. The symbols used to represent features can be of three types: points, lines, and polygons. Points are used to depict features like bridges and buildings. Lines are used to graphically illustrate features that are linear. Some common linear features include roads, railways, and rivers. However, we also need to include representations of area, in the case of forested land or cleared land; this is done through the use of color.

The set of symbols used on Canadian National Topographic System (NTS) maps has been standardized to simplify the map construction process. A description of the complete set of symbols available can be found in a published guide titled: Standards and Specifications for Polychrome Maps. This guide guarantees uniform illustration of surface features on both 1:50 000 and 1:250 000 topographic maps. Despite the existence of this guide, we can find that some topographic maps may use different symbols to depict a feature. This occurs because the symbols used are graphically refined over time - as a result the Standards and Specifications for Polychrome Maps guide is always under revision.

The tables below describe some of the common symbols used on Canadian National Topographic System maps.

Table: Transportation Features - Roads and Trails.

Feature Name	Symbol
Road - hard surface, all season	dual highway
Road - hard surface, all season	more than 2 lanes 2 lanes less than 2 lanes
Road - loose or stabilized surface, all season	2 lanes or more less than 2 lanes
Road - loose surface, dry weather	
Rapid transit route, road	
Road under construction	
Vehicle track or winter road	
Trail or portage	
Traffic circle	
Highway route number	16 401

Table: Transportation Features - Railways and Airports.

Feature Name	Symbol
Railway - multiple track	
Railway - single track	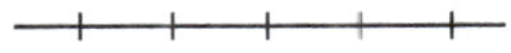
Railway sidings	
Railway - rapid transit	
Railway - under construction	
Railway - abandoned	
Railway on road	
Railway station	
Airfield; Heliport	
Airfield, position approximate	
Airfield runways; paved, unpaved	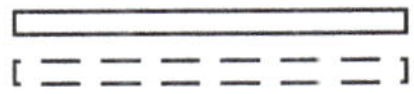

Table: Other Transportation Features - Tunnels, Bridges etc.

Feature Name	Symbol
Tunnel; railway, road	
Bridge	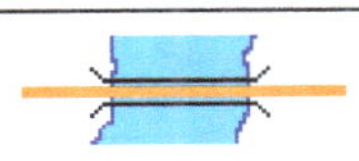
Bridge; swing, draw, lift	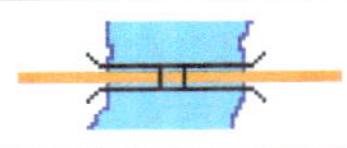
Footbridge	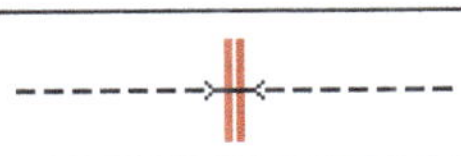
Causeway	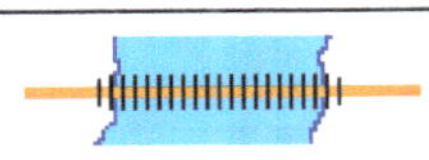
Ford	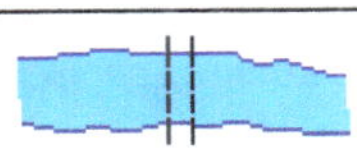

Cut	
Embankment	
Snow shed	
Barrier or gate	

Table: Hydrographic Features - Human Made.

Feature Name	Symbol
Lock	
Dam; large, small	
Dam carrying road	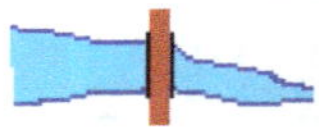
Footbridge	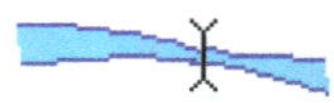
Ferry Route	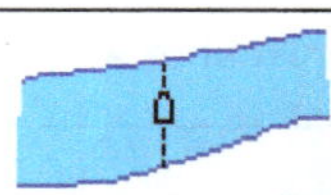
Pier; Wharf; Seawall	
Breakwater	
Slip; Boat ramp; Drydock	
Canal; navigable or irrigation	
Canal, abandoned	
Shipwreck, exposed	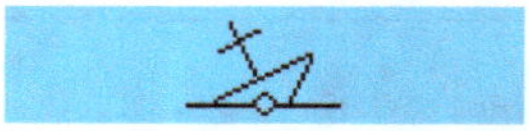
Crib or abandoned bridge pier	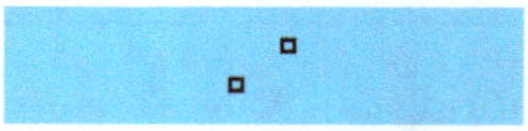
Submarine cable	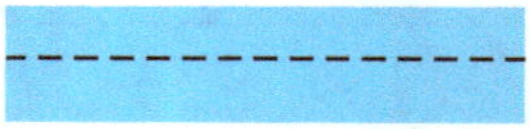
Seaplane anchorage; Seaplane base	

Table: Hydrographic Features - Naturally Occurring.

Feature Name	Symbol
Falls	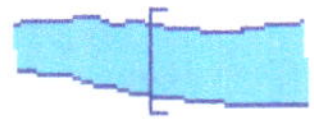
Rapids	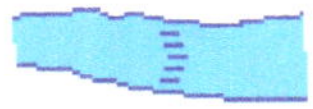
Direction of flow arrow	
Dry river bed	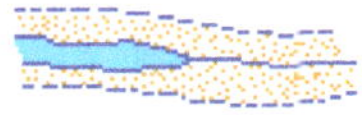
Stream - intermittent	
Sand in Water or Foreshore Flats	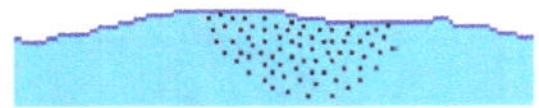
Rocky ledge, reef	
Flooded area	
Marsh, muskeg	
Swamp	
Well, water or brine; Spring	
Rocks in water or small islands	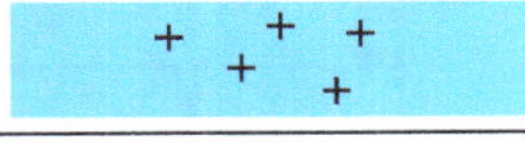
Water elevation	

Table: Terrain Features - Elevation.

Feature Name	Symbol
Horizontal control point; Bench mark with elevation	
Precise elevation	▪224
Contours; index, intermediate	
Depression contours	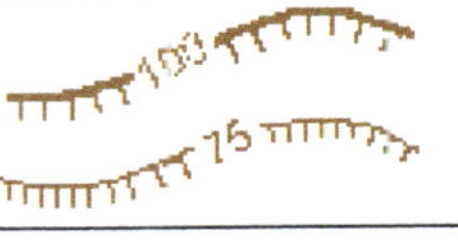

Table: Terrain Features - Geology and Geomorphology.

Feature Name	Symbol
Cliff or escarpment	
Esker	
Pingo	
Sand	
Moraine	
Quarry	
Cave	

Table: Terrain Features - Land Cover.

Feature Name	Symbol
Wooded area	
Orchard	
Vineyard	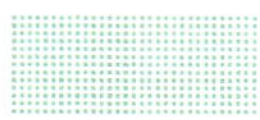

Table: Human Activity Symbols - Recreation.

Feature Name	Symbol
Sports track	
Swimming pool	
Stadium	
Golf course	

Golf driving range	
Campground; Picnic site	
Ski area, ski jump	
Rifle range with butts	
Historic site or point of interest; Navigation light	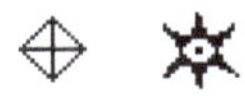
Aerial cableway, ski lift	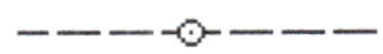

Table: Human Activity Symbols - Agriculture and Industry.

Feature Name	Symbol
Silo	
Elevator	
Greenhouse	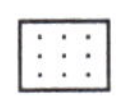
Wind-operated device; Mine	
Landmark object (with height); tower, chimney, etc.	
Oil or natural gas facility	
Pipeline, multiple pipelines, control valve	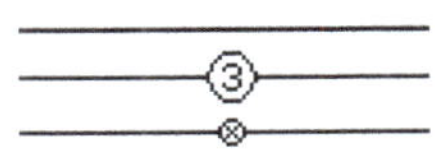
Pipeline, underground multiple pipelines, underground	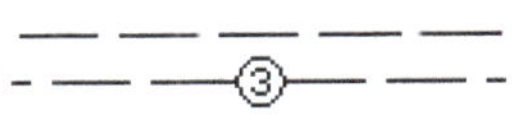
Electric facility	
Power transmission line multiple lines	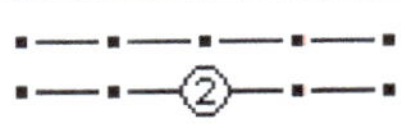
Telephone line	
Fence	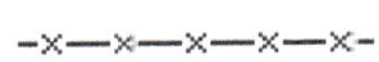

Crane, vertical and horizontal	
Dyke or levee	
Firebreak	....
Cut line	----------

Table: Human Activity Symbols - Buildings.

Feature Name	Symbol
School; Fire station; Police station	F Police
Church; Non-Christian place of worship; Shrine	+
Building	
Service centre	(S)
Customs post	
Coast Guard station	(G)
Cemetery	C
Ruins	R
Fort	

Contour Lines

Topographic maps can describe vertical information through the use of contour lines (contours). A contour line is an isoline that connects points on a map that have the same elevation. Contours are often drawn on a map at a uniform vertical distance. This distance is called the contour interval. The map in the Figure shows contour lines with an interval of 100 feet. Note that every fifth brown contour lines is drawn bold and has the appropriate elevation labeled on it. These contours are called index contours. On Figure they represent elevations of 500, 1000, 1500, 2000 feet and so on. The interval at which contours are drawn on a map depends on the amount of the relief depicted and the scale of the map.

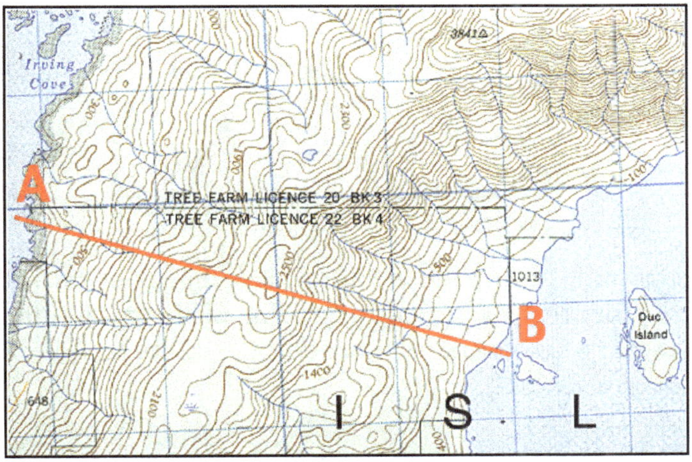

In figure portion of the "Tofino" 1:50,000 National Topographic Series of Canada map. The brown lines drawn on this map are contour lines. Each line represents a vertical increase in elevation of 100 feet. The bold brown contour lines are called index contours. The index contours are labeled with their appropriate elevation which increases at a rate of 500 feet. Note the blue line drawn to separate water from land represents an elevation of 0 feet or sea-level.

Contour lines provide us with a simple effective system for describing landscape configuration on a two-dimensional map. The arrangement, spacing, and shape of the contours provide the user of the map with some idea of what the actual topographic configuration of the land surface looks like. Contour intervals the are spaced closely together describe a steep slope. Gentle slopes are indicated by widely spaced contours. Contour lines that V upwards indicate the presence of a river valley. Ridges are shown by contours that V downwards.

Topographic Profiles

A topographic profile is a two-dimensional diagram that describes the landscape in vertical cross-section. Topographic profiles are often created from the contour information found on topographic maps. The simplest way to construct a topographic profile is to place a sheet of blank paper along a horizontal transect of interest. From the map, the elevation of the various contours is transferred on to the edge of the paper from one end of the transect to the other. Now on a sheet of graph paper use the x-axis to represent the horizontal distance covered by the transect. The y-axis is used to represent the vertical dimension and measures the change in elevation along the transect. Most people exaggerate the measure of elevation on the y-axis to make changes in relief stand out. Place the beginning of the transect as copied on the piece of paper at the intersect of the x and y-axis on the graph paper. The contour information on the paper's edge is now copied onto the piece of graph paper. Figure shows a topographic profile drawn from the information found on the transect A-B above.

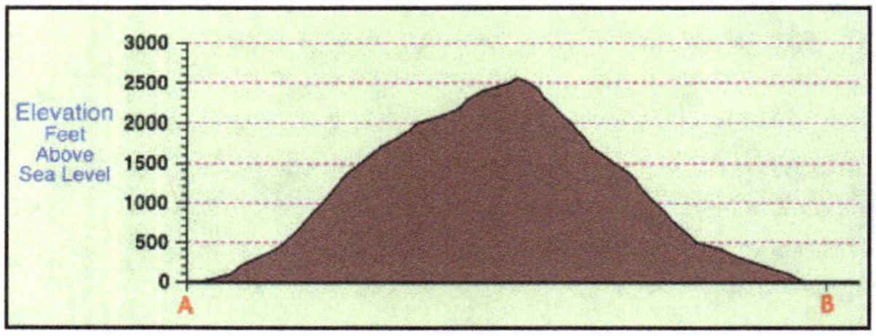

The following topographic profile shows the vertical change in surface elevation along the transect AB from Figure. A vertical exaggeration of about 4.2 times was used in the profile (horizontal scale = 1:50,000, vertical scale = 1:12,000 and vertical exaggeration = horizontal scale/vertical scale).

Military Grid Reference System and Map Location

Two rectangular grid systems are available on topographic maps for identifying the location of points. These systems are the Universal Transverse Mercator (UTM) grid system and the Military Grid Reference System. The Military Grid Reference System is a simplified form of Universal Transverse Mercator grid system and it provides a very quick and easy method of referencing a location on a topographic map. On a topographic maps with a scale 1:50,000 and larger, the Military Grid Reference System is superimposed on the surface of map as blue colored series of equally spaced horizontal and vertical lines. Identifying numbers for each of these lines is found along the map's margin. Each identifying number consists of two digits which range from a value of 00 to 99. Each individual square in the grid system represents a distance of a 1000 by 1000 meters and the total size of the grid is 100,000 by 100,000 meters.

One problem associated with the Military Grid Reference System is the fact that reference numbers must be repeated every 100,000 meters. To overcome this difficulty, a method was devised to identify each 100,000 by 100,000 meter grid with two identifying letters which are printed in blue on the border of all topographic maps (note some maps may show more than one grid). When making reference to a location with the Military Grid Reference System identifying letters are always given before the horizontal and vertical coordinate numbers.

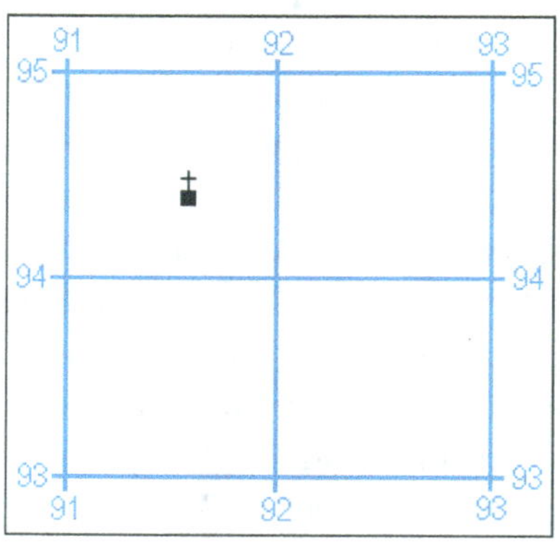

Portion of a Military Grid Reference System found on a topographic map. Coordinates on this system are based on a X (horizontal increasing from left to right) and Y (vertical increasing from bottom to top) system. The symbol depicting a church is located in the square 9194. Note that the value along the X-axis (easting) is given first followed by the value on the Y-axis (northing).

Each individual square in the Military Grid Reference System can be further divided into 100 smaller squares (ten by ten). This division allows us to calculate the location of an object to within 100 meters. This indicates that the church is six tenths of the way between lines 91 and 92 and four tenths of the way between lines 94 and 95. Using these values, we can state that the easting as being 916 and the northing as 944. By convention, these two numbers are combined into a coordinate reference of 916944.

THEMATIC MAP

A thematic map is a map that focuses on a specific theme or subject area such as physical phenomena like temperature variation, rainfall distribution and population density in an area. Thematic maps emphasize spatial variation of human issues like population density or prevalence of diseases. This is in contrast to general reference maps, which just show natural features like landforms, lines of transportation, rivers, human settlements and political and administrative boundaries. Thematic maps use the base data of general reference maps as reference points for mapping specific themes, and to enhance the understanding of the map's theme and purpose. Normally, however, all thematic maps use maps with coastlines, city locations and political boundaries as their base maps. The map's specific theme is then layered onto this base map via different mapping programs and technologies like a geographic information system (GIS). Thematic maps did not develop as a map type until the mid-17th Century because accurate base maps were not present prior to this time. The first thematic maps were created only after accurate base maps displaying coastlines, cities and other boundaries became available.

While general reference maps show where something is in space, thematic maps tell a story about that place, for example. A city map. Thematic maps are sometimes referred to as graphic essays that portray spatial variations and interrelationships of geographical distributions. Location, of course, is important to provide a reference base of where selected phenomena are occurring.

Mapping Considerations

There are several things to consider while designing thematic maps. The most significant consideration ought to be the users, since this helps to decide as to which reference points need to be included in addition to the theme. For example, a thematic map intended for use by political scientists may include political boundaries, while one intended for a geologist might need contours showing elevation. The sources of a thematic map's data are also important and should be carefully considered. In order to make the most useful maps, one must find accurate, current and reliable sources of information on a wide range of subjects – from environmental features to demographic data.

Types of Thematic Maps

There are three categories of thematic maps – univariate, bivariate and multivariate. A thematic map is univariate if the non-location data is all of the same kind. Population density, cancer rates, and annual rainfall are three examples of univariate data. Bivariate mapping shows the geographical distribution of two distinct sets of data. For example, a map showing both rainfall and elevation may be used to explore a possible correlation between the two variables. More than two sets of data leads to multivariate mapping. For example, a single map might show rainfall in addition to elevation and vegetation density in order to understand their interrelationships.

Cartographers use many methods to create thematic maps, but five techniques are especially noted:

- Choropleth Maps: A choropleth map is a thematic map in which areas are shaded or patterned in proportion to the measurement of the statistical variable being displayed on the map, such as population density or per-capita income. This technique assumes a relatively even distribution of the measured phenomenon within each region. Generally speaking, differences in hue are used to indicate qualitative differences, such as land use, while differences in saturation or lightness are used to indicate quantitative differences, such as population.

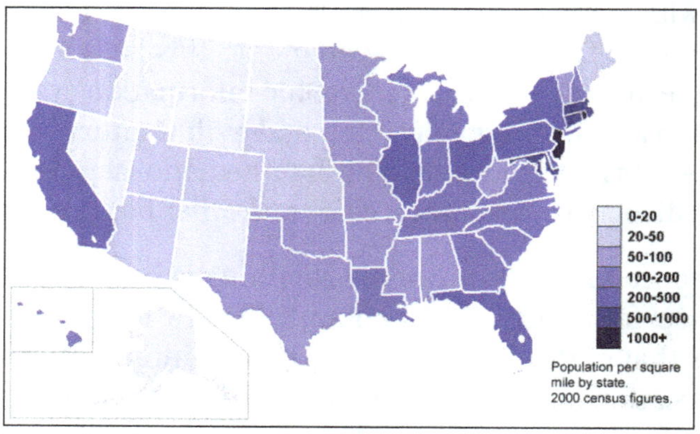

Choropleth map showing percentage of vegetarian population in American states.

- Proportional Symbol Maps: Proportional symbol maps scale the size of simple symbols (usually a circle or square) proportionally to the data value found at that location. They are a simple concept to grasp: The larger the symbol, the "more" of something exists at a location. For example, a disc may be shown at the location of each city in a map, with the area of the disc being proportional to the population of the city.

Proportional symbol map of selected American cities.

- Isarithmic or Isoline Maps: Isarithmic maps, also known as contour maps or isoline maps depict smooth continuous phenomena such as precipitation or elevation. Each line-bounded area on this type of map represents a region with the same value. For example, on an elevation map, each elevation line indicates an area at the listed elevation. An Isarithmic map is a planimetric graphic representation of a 3-D surface. Isarithmic mapping requires 3-D thinking for surfaces that vary spatially. Common isarithmic maps are of temperature, rainfall, or elevation.

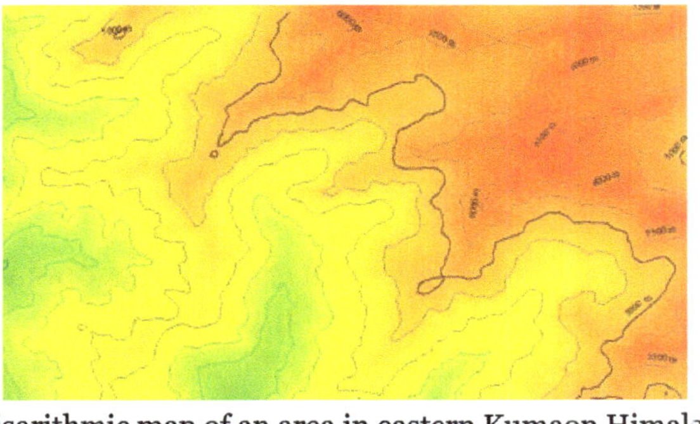

An isarithmic map of an area in eastern Kumaon Himalaya, showing elevations above mean sea level.

- Dasymetric Maps: The dasymetric map is a method of thematic mapping, which uses areal symbols to spatially classify volumetric data. Dasymetric maps are those in which densities (for example population density) are shown as they are distributed in reality, irrespective of any administrative boundaries, i.e. by natural spots of concentration and rarefaction. Map of population by areal administrative units represents a homogeneously distributed population even at places that are uninhabited, say rivers, snow cover etc. Dasymetric Mapping provides a potential solution to these problems by generation of a surface-based demographic data representation, in which data are modeled as a continuous field independent of partitioning into arbitrary areal units. Surface-based population representation offers certain advantages over areal unit representation.

- Dot Distribution Maps: A dot distribution map might be used to locate each occurrence of a phenomenon. For example, a cholera outbreak, where each dot represents one death due to cholera. Dot maps best represent data consisting of discrete observations that change more or less smoothly over space. Where appropriate, the dot distribution technique may also be used in combination with the proportional symbol technique.

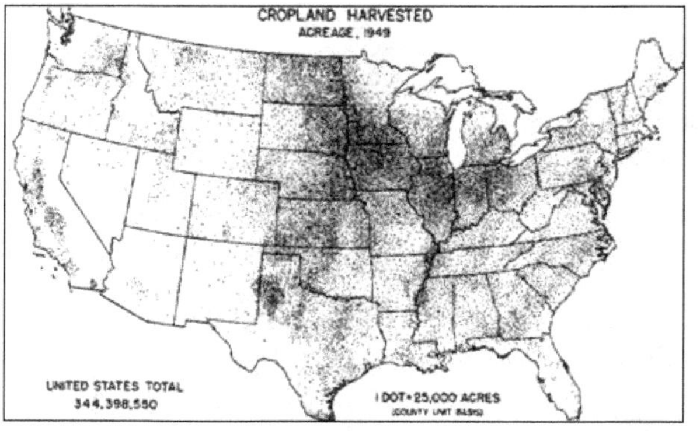

Dot distribution map showing the distribution of harvested cropland in the United States.

Uses of Thematic Maps

Thematic maps serve three primary purposes:

1. They provide specific information about particular locations.
2. They provide general information about spatial patterns.
3. They can be used to compare patterns on two or more maps.

Common examples are maps of demographic data such as population density. When designing a thematic map, cartographers must balance a number of factors in order to effectively represent the data. Besides spatial accuracy, and aesthetics, quirks of human visual perception and the presentation format must be taken into account. In addition, the audience is of equal importance. Who will "read" the thematic map and for what purpose helps define how it should be designed. A political scientist might prefer having information mapped within clearly delineated county boundaries (choropleth maps). A state biologist could certainly benefit from county boundaries being on a map, but nature seldom falls into such smooth, man-made delineations. In which case, a dasymetric map charts the desired information underneath a transparent county boundary map for easy location referencing.

City Map

A city map is a large-scale thematic map of a city (or part of a city) created to enable the fastest possible orientation in an urban space. The graphic representation of objects on a city map is therefore usually greatly simplified, and reduced to generally understood symbology.

Depending upon its target group or market, a city map will include not only the city's transport network, but also other important information, such as city sights or public institutions.

City map of Stavanger (Norway).

As early as the time of the Ancient Near East, clay tablets were being produced with scaled, graphical representations of cities. Excavations of the Sumerian city of Nippur brought to light a fragment of an approximately 3,500-year-old city map, which is sometimes referred to as the oldest known city map. The clay tablet depicts the temple of Enlil, a city park, the city wall including its gates, along with a canal and the river Euphrates. The individual objects on this map were already labelled, in a Sumerian cuneiform.

Clay tablet with map of the Babylonian city of Nippur.

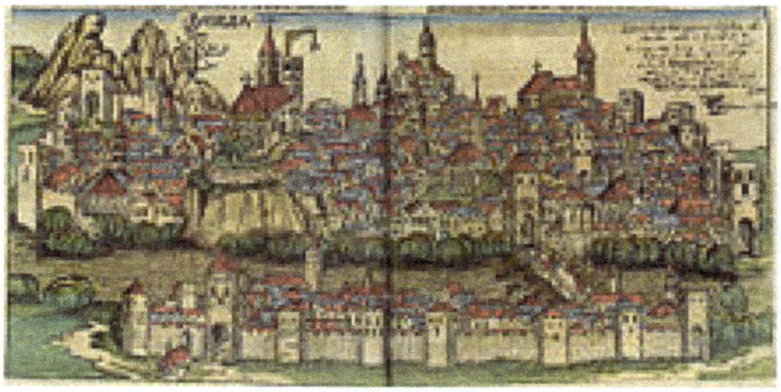

View of Basel, Switzerland, ca 1490, from the Nuremberg Chronicle.

In manuscripts and early printed books of the Late Middle Ages, cities are often shown in profile, or viewed from an elevated standpoint. Nautical charts of that time sometimes depict partly stylized cityscapes drawn in pictogram form - for example in Cristoforo Buondelmonti's *Liber insularum archipelagi* (Book of Islands), from the year 1422.

The Nuremberg Chronicle, which first appeared in 1493, is one of the most important collections of city views of the late Middle Ages, with over 100 such illustrations. Nevertheless, panoramas like this one, or the one in Bernhard von Breydenbach's*Travelogue* from 1483, had more narrative or representative functions.

Illustrated in these works are the local conditions and relevant characteristics - such as ports, magnificent buildings, walls, etc. - as a background for highlighting historical descriptions or economic benefits of the city. On the other hand, less emphasis was placed on accuracy: in the Nuremberg Chronicle, merely one quarter of the city views represented the actual appearance of the subject city, and some individual images were even used simultaneously to represent several different cities.

Antwerp (engraving), ca 1572, by Georg Braun and Frans Hogenberg.

In the 16th century, the artists and scholars of the Renaissance had extensive knowledge of mathematical perspectives and projections. This knowledge also affected the work of cartographers and the production of cityscapes (especially in Italy). A key innovation was that the city was no longer portrayed simply from an imaginary or real perspective, but drawn initially as a two-dimensional map, and then, using a process of accurate perspective drawing, transformed into a three-dimensional image.An early example of a geometrically exact and highly detailed work of this kind is the city map of Venice created by Jacopo de' Barbari in around 1500.

Whereas the illustrations of the late Middle Ages are usually still simple small-format woodcuts, an increasingly common process from 1500 onwards was the creation of prints from huge woodcuts and woodblocks. Jacopo de' Barbari's map of Venice was already as large as 139 centimetres (55 in) x 282 centimetres (111 in), and consisted of six individual panels. From the middle of the 16th century, the copperplate process, originating in Antwerp, began to compete with the woodcut, and allowed far more refined and detailed illustrations.

One of the first city pocket atlases, and the first pocket atlas of London, was "Collins' Illustrated Atlas of London" published in 1854 and drawn and engraved by Richard Jarman.

Content and Design

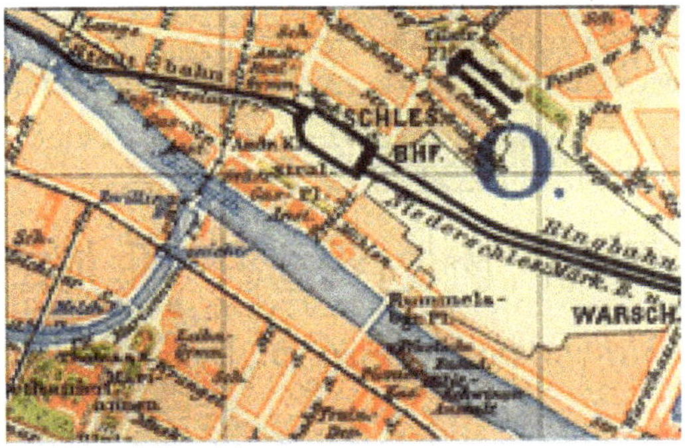

Detail from a city map of Berlin from 1895.

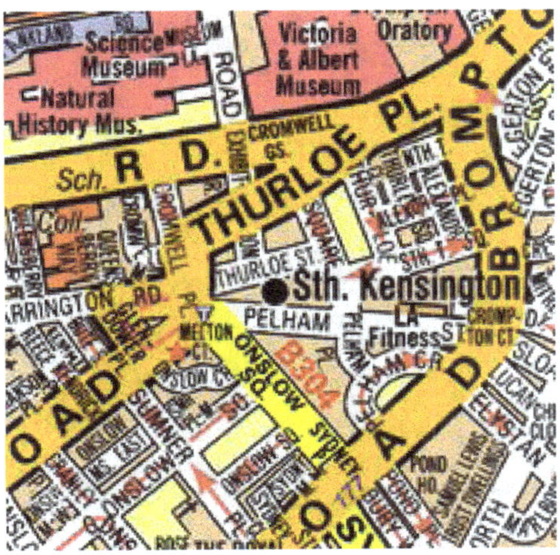

Detail of Geographers' A-Z Street Atlas of London.

The scale of a city map is usually between 1:10,000 and 1:25,000. Densely settled downtown areas will sometimes be partly drawn in a larger scale, on a separate detail map.

In addition to linear true to scale maps, there are also maps with variable scale, for example where the scale gradually increases towards the city centre (aerial photography, and photogrammetry methods).

Central to the information provided by a city map is the street network, including its street names (often supplemented by at least a selection of individual house numbers), along with buildings, parks and waterways. Streets and points of interest are usually also listed in a legend or register, locating objects on a map grid on the map. Important places such as administrative buildings, cultural institutions, attractions, etc. may be highlighted with the assistance of pictograms. The map may also be complemented by representations of public transport facilities.

Choropleth Map

A choropleth map is a thematic map in which areas are shaded or patterned in proportion to the measurement of the statistical variable being displayed on the map, such as population density or per-capita income.

Choropleth maps provide an easy way to visualize how a measurement varies across a geographic area or show the level of variability within a region. A heat map is similar but does not use geographic boundaries.

The earliest known choropleth map was created in 1826 by Baron Pierre Charles Dupin. They were first called "*cartes teintées*" (*coloured map* in French). The term "choroplethe map" was introduced in 1938 by the geographer John Kirtland Wright in "Problems in Population Mapping".

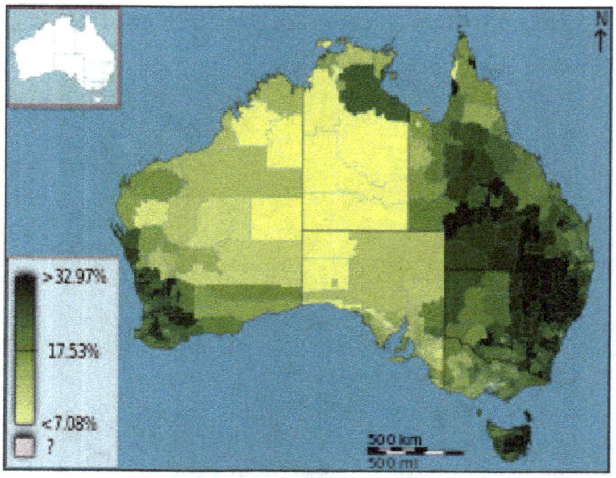

A choropleth map that visualizes the fraction of Australians that identified as Anglican at the 2011 census.

Choropleth maps are based on statistical data aggregated over previously defined regions (e.g., counties), in contrast to area-class and isarithmic maps, in which region boundaries are defined by data patterns. Thus, where defined regions are important to a discussion, as in an election map divided by electoral regions, choropleths are preferred.

Where real-world patterns may not conform to the regions discussed, issues such as the ecological fallacy and the modifiable areal unit problem (MAUP) can lead to major misinterpretations, and other techniques are preferable. Similarly, the size and specificity of the displayed regions depend on the variable being represented. While the use of smaller and more specific regions can decrease the risk of ecological fallacy and MAUP, it can cause the map to appear to be more complicated. Although representing specific data in large regions can be misleading, it can make the map clearer and easier to interpret and remember. The choice of regions will ultimately depend on the map's intended audience and purpose.

The dasymetric technique can be thought of as a compromise approach in many situations. Broadly speaking, choropleths represent two types of data: spatially extensive or spatially intensive.

- Spatially extensive data are things like populations. The population of the UK might be 65 million, but it would not be accurate to arbitrarily cut the UK into two halves of equal area and say that the population of each half of the UK is 32.5 million.
- Spatially intensive data are things like rates, densities and proportions, which can be thought of conceptually as field data that is averaged over an area. Though the UK's 60 million inhabitants occupy an area of about 240,000 km², and the population density is therefore about 250/km², arbitrary halves of equal area would not both have the same population density.

Normalization

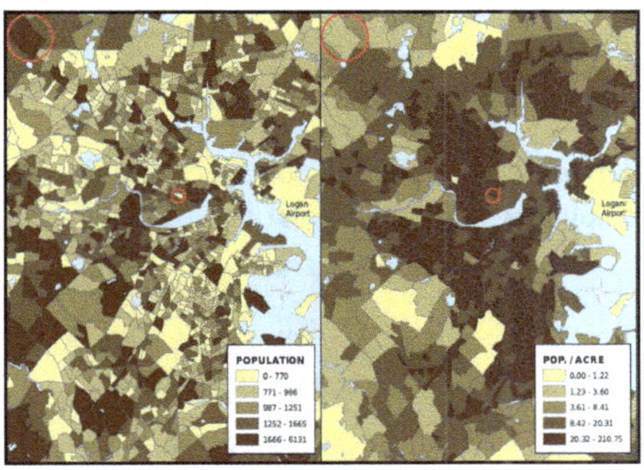

Normalization: the map on the left uses total population to determine color. This causes larger polygons to appear to be more urbanized than the smaller dense urban areas of Boston, Massachusetts. The map on the right uses population density. A properly normalized map will show variables independent of the size of the polygons.

Another common error in choropleths is the use of raw data values to represent magnitude rather than normalized values to produce a map of densities. This is problematic because the eye naturally integrates over areas of the same color, giving undue prominence to larger polygons of moderate magnitude and minimizing the significance of smaller polygons with high magnitudes. The problem with using data in total counts arises when the polygons are not all the same size (in area or total population), as in the figure at right. Because a single color, representing a single value, is spread over the entire area of the district, large areas will be more dominant in the visual hierarchy than they should be, and are commonly misinterpreted as having larger values than smaller districts with the same color. To solve this issue, one can *normalize* the variable by dividing it by the total area, thus deriving *density*, which is a field. Another solution is to represent total amounts using a proportional symbol map.

Other valid forms of normalization for choropleth maps can be derived by computing ratios between two total amounts, such as rates of change (population growth = 2010 population / 2000 population) and mean allocations (mean family income = total income / total families), or other descriptive statistics such as the median or standard deviation.

Classification

- Equal intervals
- Equal frequency
- Geometric progressions
- Standard deviation
- Jenks natural breaks optimization
- Head/tail Breaks

Color Progression

When mapping quantitative data, a specific color progression should be used to depict the data properly. There are several different types of color progressions used by cartographers.

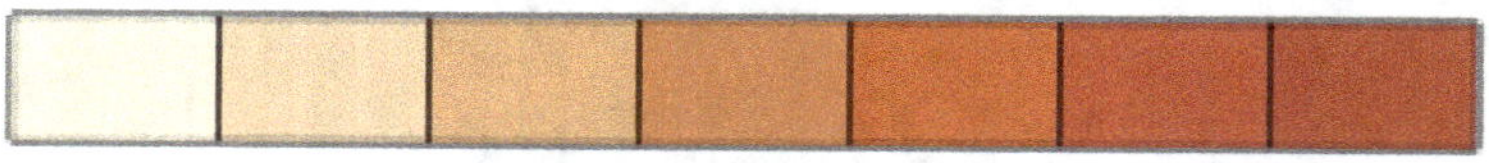

Single hue progression.

Single-hue progressions fade from a dark shade of the chosen color to a very light or white shade of relatively the same hue. This is a common method used to map magnitude. The darkest hue represents the greatest number in the data set and the lightest shade representing the least number.

Two variables may be shown through the use of two overprinted single color scales. The hues typically used are from red to white for the first data set and blue to white for the second, they are then overprinted to produce varying hues. These type of maps show the magnitude of the values in relation to each other.

Bi-polar color progression.

Bi-polar progressions are normally used with two opposite hues to show a change in value from negative to positive or on either side of some either central tendency, such as the mean of the variable being mapped or other significant value like room temperature. For example, a typical progression when mapping temperatures is from dark blue (for cold) to dark red (for hot) with white in the middle. When one extreme can be considered better than the other (as in this map of life expectancy) then it is common to denote the poor alternative with shades of red, and the good alternative with green.

Complementary hue progressions are a type of bi-polar progression. This can be done with any of the complementary colors and will fade from each of the darker end point hues into a gray shade representing the middle. An example would be using blue and yellow as the two end points.

Blended hue color progression.

Blended hue progressions use related hues to blend together the two end point hues. This type of color progression is typically used to show elevation changes. For example, from yellow through orange to brown.

Partial spectral color progression.

Partial spectral hue progressions are used to map mixtures of two distinct sets of data. This type of hue progression will blend two adjacent opponent hues and show the magnitude of the mixing data classes.

Full-spectral color progression.

Full spectral progression contains hues from blue through red. This is common on relief maps and modern weather maps. This type of progression is not recommended under other circumstances because certain color connotations can confuse the map user.

Value progression.

Value progression maps are monochromatic. Although any color may be used, the archetype is from black to white with intervening shades of gray that represent magnitude. According to Robinson *et al.* this is the best way to portray a magnitude message to the map audience. It is clearly understood by the user and easy to produce in print.

Qualitative color progression.

A Qualitative progression is often used when working with nominal or qualitative data. The colors shown on the map seem unrelated to one another or are arbitrarily chosen. For example, a choropleth map of "most prevalent religion" would be best shown with this type of scheme.

Usability

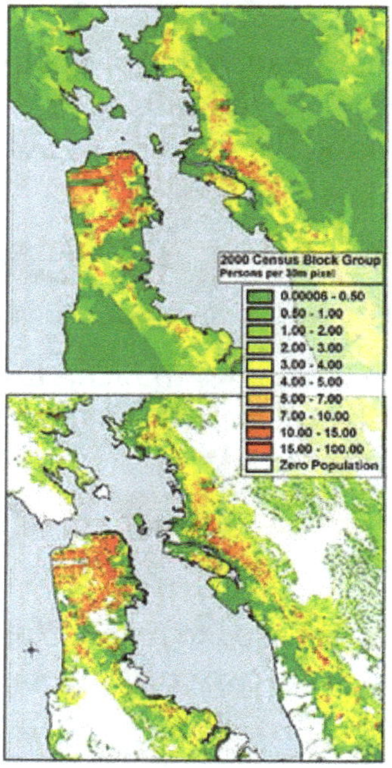

A choropleth map (top) and a dasymetric map (bottom) of the population of the San Francisco Bay Area in 2000.

When using any of these methods, there are two important principles: first is that darker colors are perceived as being higher in magnitude; second is that, while there are millions of color variations, the human eye is limited as to how many colors it can easily distinguish. Generally, five to seven color categories are recommended. The map user should be able to easily identify the implied magnitude of the hue and to match it with the legend.

Additional considerations include color blindness and various reproduction techniques. For example, the red–green bi-polar progression described in the section above is likely to cause problems for dichromats. A related issue is that color scales which rely primarily on hue with insufficient variation in saturation or intensity may be compromised if reproduced in black and white. Conversely, if a map is legible in black and white (lightness-based), then a prospective user's perception of color is irrelevant.

Color can greatly enhance the communication between the cartographer and their audience, but poor color choice can result in a map that is neither effective nor appealing to the map user. A simpler correlation between color-properties and underlying values is better.

Cartogram

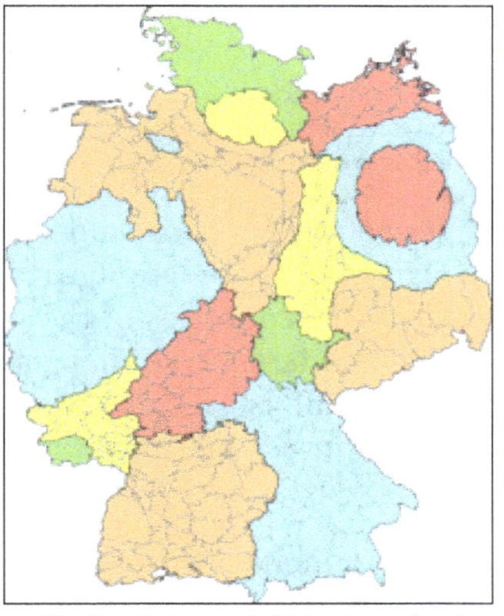

Cartogram of Germany, with the states and districts resized according to population.

A cartogram is a map in which some thematic mapping variable – such as travel time, population, or GNP – is substituted for land area or distance. The geometry or space of the map is distorted, sometimes extremely, in order to convey the information of this alternate variable. They are primarily used to display emphasis and for analysis as nomographs.

Two common types of cartograms are area and distance cartograms. Cartograms have a fairly long history, with examples from the mid-1800s.

Area Cartograms

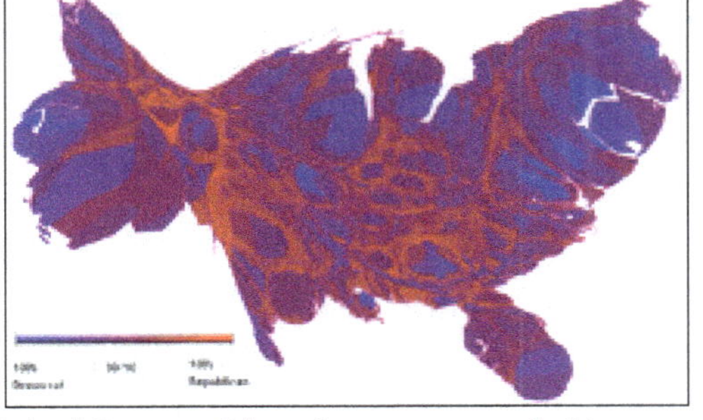

Area cartogram of the United States, with each county rescaled in proportion to its population. Colors refer to the results of the 2004 U.S. presidential election popular vote.

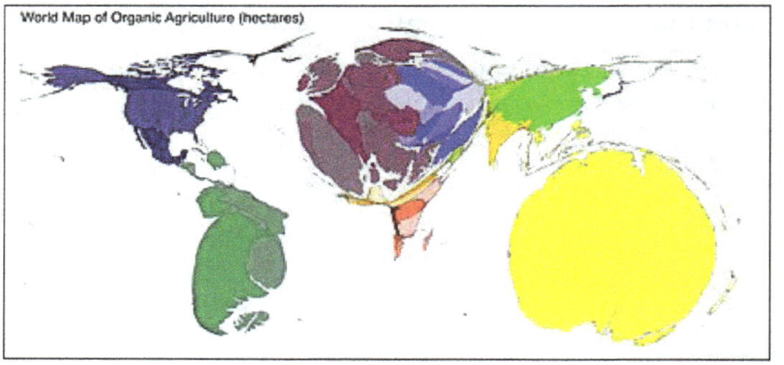

Area cartogram of the world with each country rescaled in proportion to the hectares of certified organic farming.

An area cartogram is sometimes referred to as a value-by-area map or an isodemographic map, the latter particularly for a population cartogram, which illustrates the relative sizes of the populations of the countries of the world by scaling the area of each country in proportion to its population; the shape and relative location of each country is retained to as large an extent as possible, but inevitably a large amount of distortion results. Other synonyms in use are anamorphic map, density-equalizing map and Gastner map.

Area cartograms may be contiguous or noncontiguous. The area cartograms are all contiguous, while a good example of a noncontiguous cartogram was published in *The New York Times*. This method of cartogram creation is sometimes referred to as *the projector method* or *scaled-down regions*.

Cartograms may be classified also by the properties of shape and topology preservation. Classical area cartograms are typically distorting the shape of spatial units to some degree, but they are strict at preserving correct neighborhood relationships between them. Scaled-down cartograms (from the NY Times example) are strictly shape-preserving. Another branch of cartograms introduced by Dorling, replaces actual shapes with circles scaled according to the mapped feature. Circles are distributed to resemble the original topology. Demers cartogram is a variation of Dorling cartogram, but it uses rectangles instead of circles, and attempts to retain visual cues at the expense of minimum distance. Schematic maps based on quad trees can be seen as non shape-preserving cartograms with some degree of neighborhood preservation.

Production

One of the first cartographers to generate cartograms with the aid of computer visualization was Waldo Tobler of UC Santa Barbara in the 1960s. Prior to Tobler's work, cartograms were created by hand (as they occasionally still are). The National Center for Geographic Information and Analysis located on the UCSB campus maintains an online Cartogram Central with resources regarding cartograms.

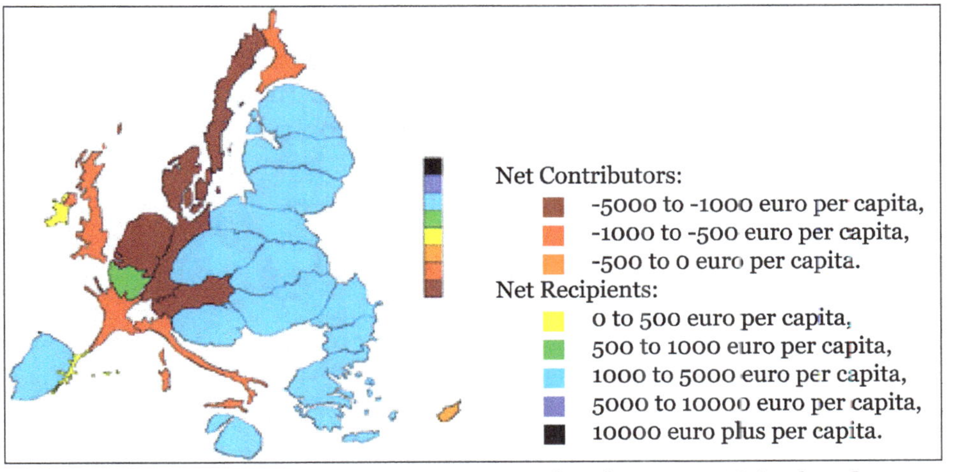

Cartogram showing Open Europeestimate of total European Union based on Eurostat 2007 pop. estimates (Luxembourg not shown).

A number of software packages generate cartograms. Most of the available cartogram generation tools work in conjunction with other GIS software tools as add-ons or independently produce cartographic outputs from GIS data formatted to work with commonly used GIS products. Examples of cartogram software include ScapeToad, Cart, and the Cartogram Processing Tool (an ArcScript for ESRI's ArcGIS), which all use the Gastner-Newman algorithm. An alternative algorithm, Carto3F, is also implemented as an independent program for non-commercial use on Windows platforms. This program also provides an optimization to the original Dougenik rubber-sheet algorithm. The CRAN package recmap provides an implementation of a rectangular cartogram algorithm.

Cartograms can also be constructed manually, either by hand or in a computer-assisted environment. Block cartograms are constructed by arranging geometrically regular equal-sized blocks, with the number of blocks allocated to each district proportional to the population variable. Several examples of block cartograms were published during the 2016 U.S. presidential election season by *The Washington Post*, the *FiveThirtyEight* blog, and the *Wall Street Journal*, among others.

Flow Map

Flow maps are a type of thematic map used in cartography to show the movement of objects between different areas. These types of maps can show things like the movement of goods across space, the number of animal species in a specific migration pattern, as well as traffic volume and stream flow. They can also show both qualitative and quantitative data. Flow maps usually represent the movement of goods, weather phenomena, people and other living things with line symbols of different widths. Thus, the use of lines on a flow map is similar to the use of graduated symbols on other types of thematic maps.

When properly designed, flow maps are beneficial because they allow cartographers, GIS analysts and map users alike to easily see the differences in magnitude of a wide variety of items across space with very little map clutter. This in turn allows businesses to see where the majority of their products are going, commuters to see traffic patterns, and meteorologists to see wind patterns.

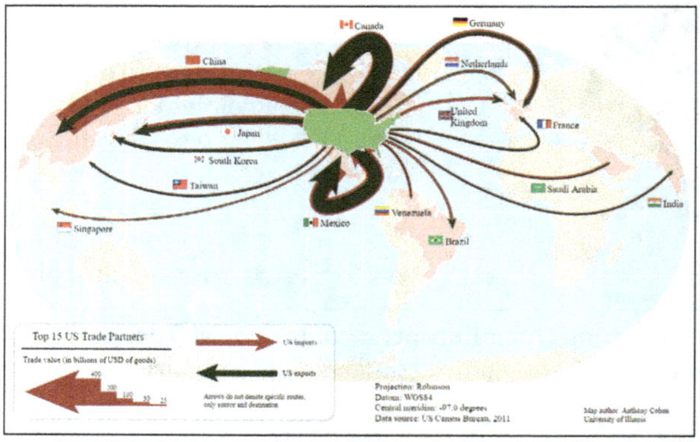

Flow map of us imports and exports to its top 15 trade partners.

How Flow Maps Work

Flow maps typically use lines to show the movement of people and goods between various locations. The lines are varied in width to represent the quantity of flow (sathyaprasad). Therefore if there is a very wide line showing traffic on one california roadway and a very thin line showing traffic on another, the road with the wider line is generally the one that contains more traffic.

Because the movement of goods and people are usually shown lines, most flow maps are created with vector, instead of raster, based data so that they can show movement continuously over the earth's surface (buckley). In vector based flow maps, the vectors are points or lines that hold information about the direction and magnitude of item that is moving (buckley). The points and lines can then be overlaid onto a map to show the movement throughout a given area.

Vectors can be symbolized in a flow map with different orientations, point size, and line length or width to show direction and magnitude. For example, flow maps showing global wind patterns often have lines with arrows on them. The point of the arrow shows the direction of movement (straight, circular, curved, etc.), while the width of the line shows the wind's intensity.

Although most flow maps use vector data, ArcGIS has recently introduced a tool for distributive flow maps that uses raster data. The distributive flow lines tool (DFLT) is a spatial analyst tool in arcgis that generates distributive flow lines from one source to many different destination points. The DFLT is completely raster based until the end when it creates a vector based flow line feature class to be used on flow maps (Bgerit).

Some arcgis users say this raster based tool is optimal because it allows for more control over the flow lines and it decreases processing time (Bgerit).

Another important thing to note about flow maps is that they can use and display both qualitative and quantitative data. For qualitative data the maps usually display symbols of uniform width that just show movement with arrows (mcgraw-hill). This data is a connection of some sort and it is not based on magnitude. Quantitative flow mapping uses lines and symbols of different widths and sizes to show changes in magnitude between areas.

Types of Flow Maps

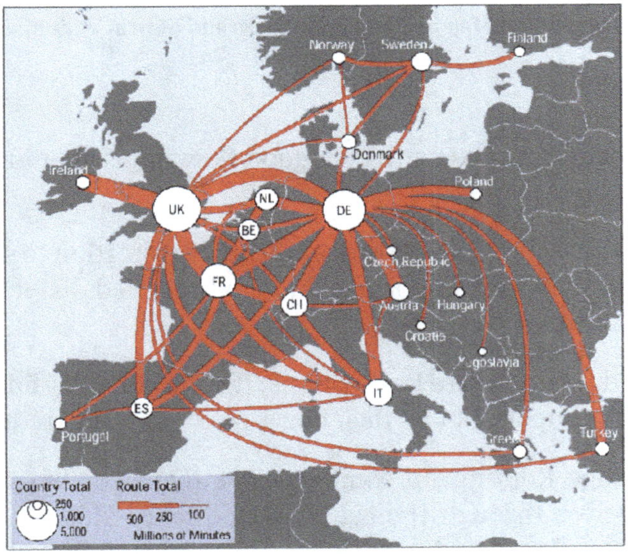

Telecommunications traffic flow map, 2000.

When looking at and creating flow maps it is important to note that there are three basic categories for the maps. These are radial, network and distributive. Radial flow maps show relationships between one source and many destinations and use separate lines radiating out from a starting point to show movement. Network flow maps show the quantity of flow over an existing network (sathyaprasad). These types of flow maps most frequently show transportation and communication networks. Distributive flow maps are maps that show relationships between a single source and many destinations like a radial flow map. These maps are different however, because they often have a large, single line produced from one source and that forks into many smaller lines once they reach their destination.

In addition to these three map categories, flow maps can also be still or animated. The traditional printed wind, traffic and other maps showing the movement of goods across space are still but computers now allow interactive, animated flow maps that can show things like wind speed during and after a hurricane for instance.

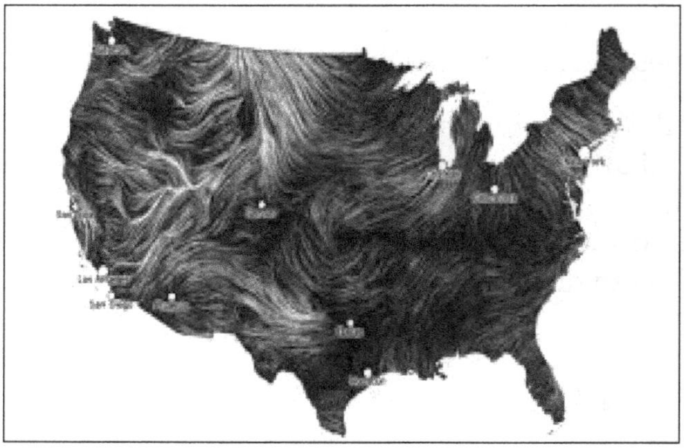

Dynamic Wind Map by Fernanda Viégas and Martin Wattenberg.

Good Flow Map

Whether radial, network or distributive all good flow maps should have the following characteristics and components:

- Intelligent Distortion: Some flow maps feature distortion to show the movement of goods. Therefore it is important that any intended distortion not change the meaning of the map.
- Merging of Edges that Share Destinations: If there are many lines going to the same destination it is important that their edges be combined to reduce map clutter.
- Intelligent Edge Routing: In some cases branches or lines on flow maps will route themselves through the center of the map. This can obscure the other lines so they can be routed to the edge of the map so that all of the data can be easily seen.
- Layering and Branching Structure: Some flow maps have a common set of nodes. In these cases layering of their lines works well to reduce map clutter.
- Linear or Logarithmic Display Widths: Flow maps can use both linear and logarithmic display widths. It is important to choose the correct one to best show the data.

Examples of Flow Mapping

Because flow maps can use a wide variety of data there are also many different projects in a plethora of fields that can use this mapping technique. Business geographers can for example, use flow mapping to examine the amount of coal exports from a specific country. Hydrologists can monitor stream flow for a particular state or region. City planners and transportation geographers can use flow mapping to examine traffic patterns and the volume of cars on specific roadways to determine the best places for new businesses and residential developments.

Map of voting cardinals showing where they arrived to the vatican from.

Dasymetric Map

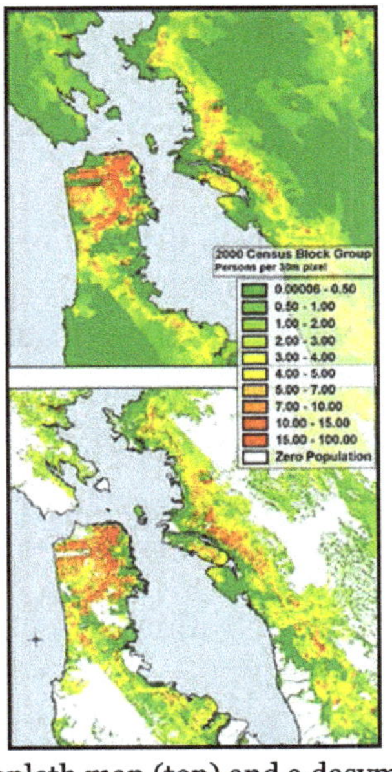

Comparison of a choropleth map (top) and a dasymetric map (bottom) of the population of the San Francisco Bay Area in 2000.

The dasymetric map is a method of thematic mapping, which uses areal symbols to spatially classify volumetric data. The method was defined and developed in 1911 by Benjamin (Veniamin) Petrovich Semenov-Tyan-Shansky and popularised by J.K. Wright, although there are earlier references to similar techniques from George Poulett Scrope and Henry Drury Harness.

Mapping

Semenov-Tyan-Shansky defined dasymetric maps as maps "on which population density, irrespective of any administrative boundaries, is shown as it is distributed in reality, i.e. by natural spots of concentration and rarefaction." Cartographers use dasymetric mapping for population density over other methods because of its ability to realistically place data over geography. Considered a hybrid or compromise between isopleth and choropleth maps, a dasymetric map utilizes standardized data, but places areal symbols by taking into consideration actual changing densities within the boundaries of the map. To do this, ancillary information is acquired, which means the cartographer steps statistical data according to extra information collected within the boundary. If appropriately executed, it is far superior to choropleth maps in relaying statistical data within areas of interest.

Like other forms of thematic mapping, the dasymetric method was created and historically used because of the need for accurate visualization methods of population data. Dasymetric maps are not widely used because of the limited options for producing them with automated tools such as geographic information systems. Although fields such as public health still rely on choropleth maps, dasymetric maps are becoming more prevalent in developing fields, such as conservation and sustainable development.

Isoline Maps

The isoline representation is the most used method to visualise quantitative phenomena which occur comprehensively and which values vary continuously in space. They are therefore called continua. Examples for such continua are temperature, air pressure, precipitation heights or ground elevations.

Isolines are lines which connect points with identical values inside a continuum. Isolines are virtual and abstract. Gradients are related to isolines and show the direction of the biggest value differences at a specific point. Gradients are always perpendicular to isolines.

The map below shows an example of an isoline map.

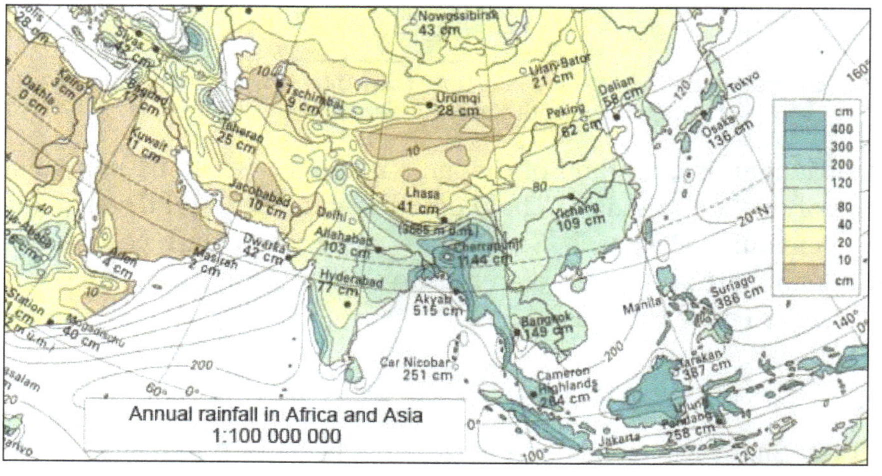

Annual rainfall in Africa and Asia
1:100 000 000

Natural and Geometrical Continua

Isolines of natural continua include in particular geophysical, geochemical and other continuous natural phenomena. In the fields of climate and meteorology isoline maps are often used. There are particular names for isolines of specific continua. Here are some examples:

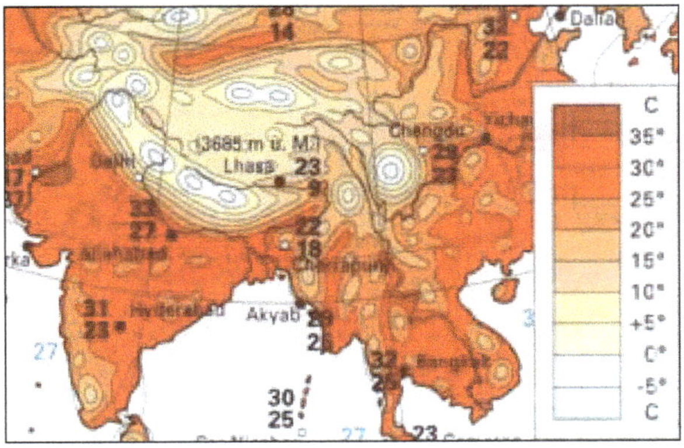

Isotherme: lines of same temperatures.

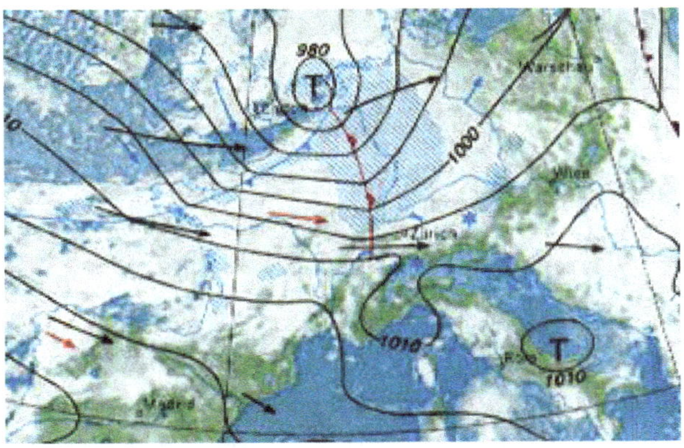

Isobar: lines of same air pressures.

Isolines of geometrical continua do not exist naturally. They are calculated or constructed. Examples of geometrical continua are:

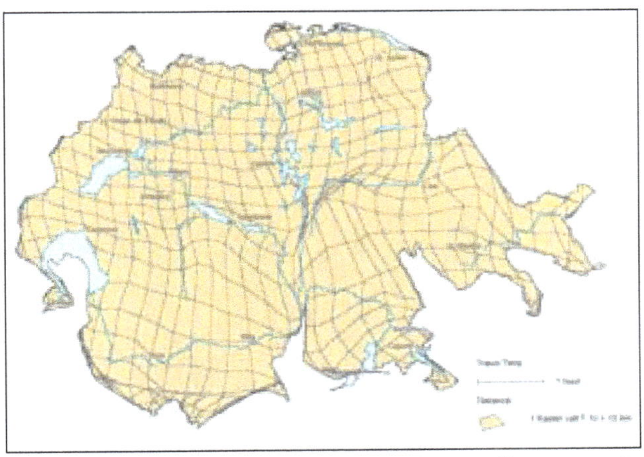

Isodistants: line of same distant to a reference line or a point.

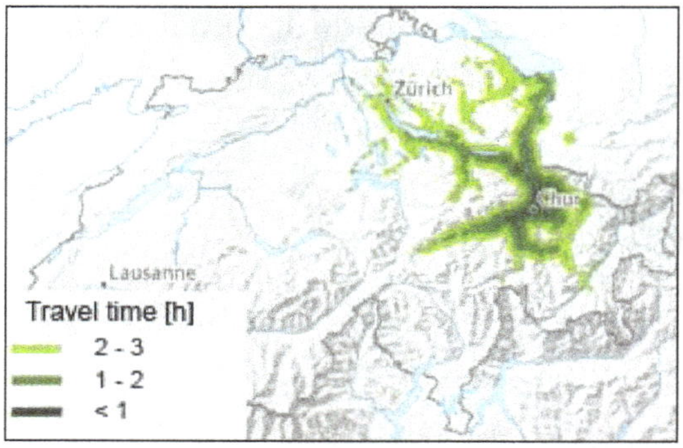

Distortion isograms: lines of same distortion.

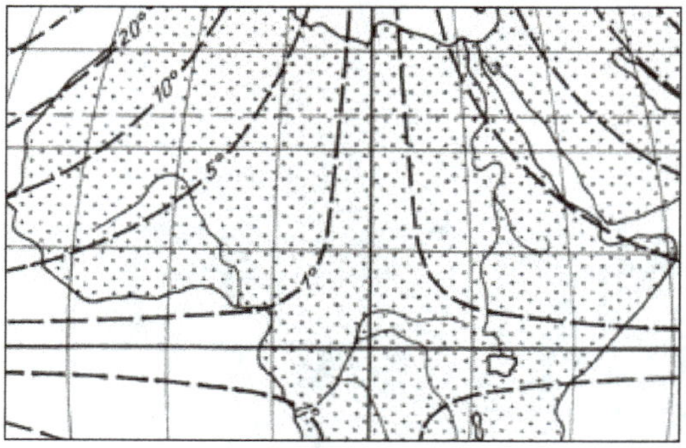

Design forms of Isolines

Value Gradations

For visualisations with isolines mostly equidistant gradations are used. Depending on the topic or the data also other forms of gradations are possible. For contour lines in areas like Switzerland, which include steep as well as flat regions, a combination of two equidistant gradations are suitable.

The left image shows a steep area in the Valais alps. To reach a good readability an equidistance of 20m is used.

Landeskarte 1:25000, reproduced with permission from swisstopo.

For the more flat area in the Swiss midland a smaller equidistance of 10m is more suitable.

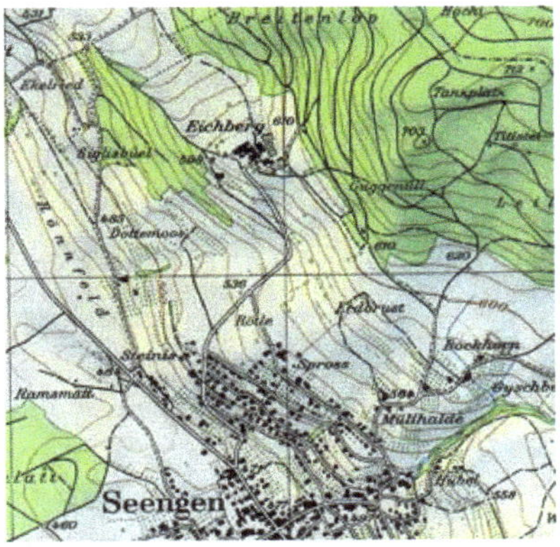

Landeskarte 1:25000, reproduced with permission from swisstopo.

Besides this possibility also progressively increasing or random gradation can be used, whereby the second one is not recommended.

Fill Colours

To increase the readability different levels of values can be summarised and coloured with a fill. The colours should be selected regarding the visualised topic. E.g. warm and cold regions could be coloured with reddish and blueish colours respectively.

Labelling

If there is an isoline map with different fill colours between the isolines there should be an explanation in the legend.

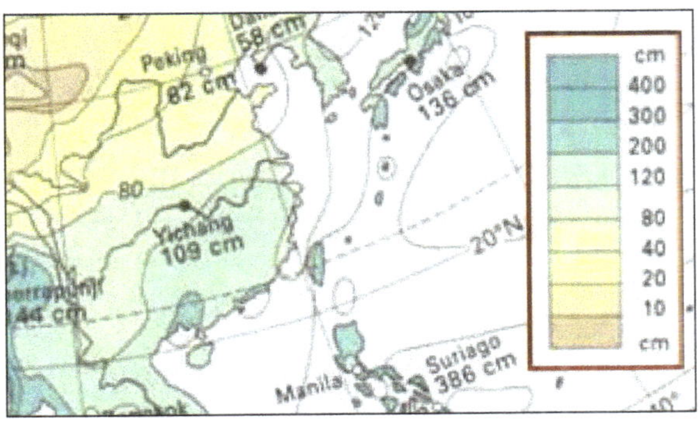

Isolines without fill colours are labelled directly in the map with their corresponding values. Depending on its length and geometry a isoline can by no means have multiple labels. This prevents the map reader from laborious search for the label. If there is not enough space to label each line, only main isolines can be labelled (e.g. only 100 or 1000). These main isoline should also be emphasized with a slightly greater line width.

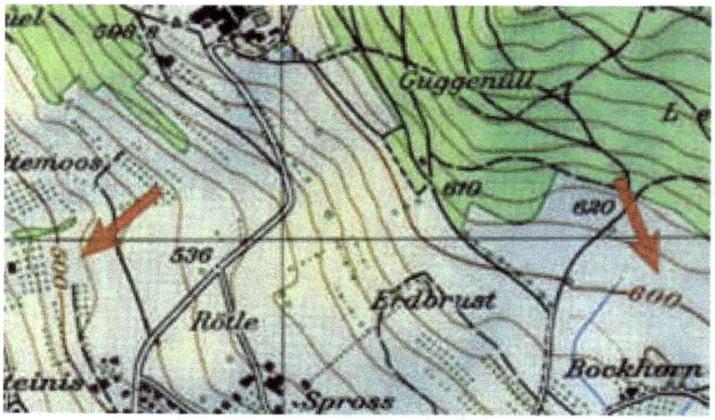

Landeskarte 1:25000, reproduced with permission from swisstopo.

Isoline maps help the reader to recognise patterns and relationships between the geography of an area and data that might have been collected on the ground, such as air temperature.

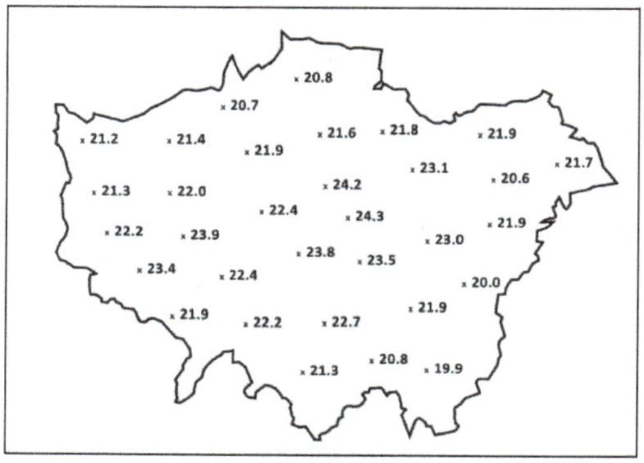

An outline map of Greater London showing the maximum temperature (in °C) recorded on a given day in the summer.

Isolines are lines drawn on a map connecting data points of the same value. They are commonly used by geographers. Contour lines, for example, show relief and connect points on the map that have the same height. Equally, isobars show bands of high and low pressure and connect points that have the same atmospheric pressure.

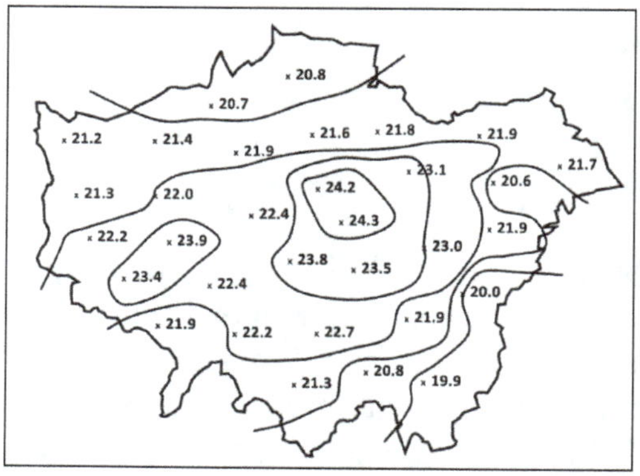

The same map with appropriate isolines used (at 1°C intervals).

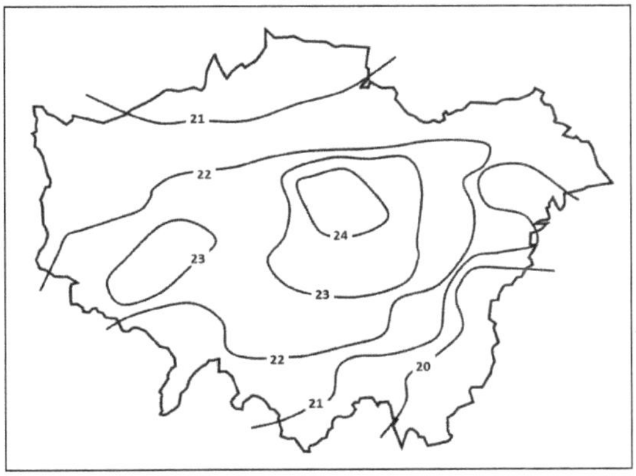

This map can be further simplified by removing the actual data points and only showing the isolines with their labels.

Isolines should have equal intervals between them numerically. The scale used (for example, whether the value goes up in tens or hundreds) depends on the nature of the data being used and at which scale the map will tell the reader the most information. As equal intervals are used, it is unlikely that an isoline will actually pass through every point that has been plotted, instead passing either side of the point depending on whether the value of the isoline itself represents a higher or lower value than the data point.

If the areas between the isolines are shaded in a choropleth fashion, the graphic is known as an Isopleth Map.

References

- Map, science: britannica.com, Retrieved 19 July, 2019
- Introduction-geologic?qt-science-center-objects=0qt-science-center-objects, science, national-cooperative-geologic-mapping-program, core-science-systems: usgs.gov, Retrieved 2 April, 2019
- Whats-in-a-map: gislounge.com, Retrieved 25 May, 2019
- “kulturelles erbe köln: johns, jasper, map (weltkarte)”. Www.kulturelles-erbe-koeln.de. Retrieved 2017-01-18
- Reference-maps: mapsofindia.com, Retrieved 29 April, 2019
- Nippur - sacred city of enlil. Oriental institute of the university of chicago. Retrieved on 24 may 2010
- Fundamentals: physicalgeography.net, Retrieved 30 June, 2019
- Dent, borden; torguson, jeffrey; hodler, thomas. Cartography thematic map design. Mcgraw-hill. Isbn 978-0-072-94382-5
- Overview-flow-mapping: gislounge.com, Retrieved 19 August, 2019
- Petrov, andrey (2012). “one hundred years of dasymetric mapping: back to the origin”. The cartographic journal. 49 (3): 256–264. Doi:10.1179/1743277412y.0000000001
- Typogrdesign-learningobject7, thematiccart: gitta.info, Retrieved 10 May 2019

4

Web Cartography and Web-enabled GIS

Web cartography, also known as web mapping, is the process of using maps delivered by GIS on the World Wide Web. Some of the well-known web maps are Yahoo maps, Bing maps and Google maps. This chapter closely examines these key web maps to provide an extensive understanding of the web cartography and web-enabled GIS.

WEB MAPPING

Web mapping is the process of designing, implementing, generating and delivering maps on the World Wide Web. While web mapping primarily deals with technological issues, web cartography additionally studies theoretic aspects: the use of web maps, the evaluation and optimization of techniques and workflows, the usability of web maps, social aspects, and more. Web GIS or Internet GIS is related to web mapping but with an emphasis on analysis, processing of project specific geodata as well as exploratory aspects. Often the terms web GIS and web mapping are used synonymously, even if they don't mean the same. In fact the boundary between web maps and web GIS is blurry. Web maps are often a presentation media in web GIS and web maps are increasingly gaining analytical capabilities.

A special case of web maps are mobile maps, displayed on mobile computing devices, such as mobile phones, smart phones, PDAs, GPS and other devices. If the maps on these devices are displayed by a mobile web browser or web user agent, they can be regarded as mobile web maps. If the mobile web maps also display context and location sensitive information, such as points of interest, the term location based services is frequently used.

The use of the web as a dissemination medium for maps can be regarded as a major advancement in cartography and opens many new opportunities, such as real-time maps, cheaper dissemination, more frequent and cheaper updates of data and software, personalized map content, distributed data sources and sharing of geographic information. It also implicates many challenges due to technical restrictions (low display resolution and limited bandwidth, in particular with small devices), copyright and security issues, reliability issues and technical complexity. While the first web maps were primarily static, due to technical restrictions, today's web maps can be fully interactive and integrate multiple media. This means that both web mapping and web cartography also have to deal with interactivity, usability and multimedia issues.

Naturally, the history of web maps is closely tied to the history and technological advancements of the web. Until the advent of the WWW, invented at CERN in 1989 and released to the public in 1993, the data transferred over the internet was primarily based on text or proprietary data formats. It is not clearly documented when the first map was published on the web, but shortly after HTML became a standard, in June 1993, Xerox Parc already implemented the first interactive map server on the web. This CGI map server implementation written in Perl allowed the user to specify a map extent, choose map layers, a map projection and some styling parameters. The response was a HTML file and an embedded gif raster image representing the map. It can be regarded as the predecessor of today's standardized WMS (Web Map Service). In 1994, the National Atlas of Canada went online as the first major online atlas worldwide. Later that year, the first version of the popular Netscape browser was released.

In March 1995, Sun made the first version of Java available, and was introduced in the main web browsers soon after. Java Applets are a huge step forward regarding interactive, platform independent applications on the web. From the beginning in 1995 until now, Java applets were used to deliver interactive map content on the web. In December 1995, Brendan Eich (Netscape) introduced the first version of Java Script which enabled first interactive websites, including simple interactivity in raster based maps.

Early in 1996, Mapquest, one of the first popular address matching and routing services went online. Results were displayed using web maps and textual descriptions. In the middle of 1996, Multimap, a UK based web mapping service went online. It is now one of the UKs most popular websites. In late 1996, Geomedia Web Map 1.0 was published, one of the first commercial web map server. At about the same time, Macromedia released the Flash Player 1.0, after buying the technology from Future wave. Gradually, Flash grew popular for multimedia and animation content on the web, but it took a couple years (until about 2000), until Flash was suitable for more serious applications with a mature scripting language available. In 1997, USGS received the mandate to coordinate and create the Online National Atlas of the United States.

In the middle of 1997, UMN Map Server 1.0 was released. It grew into the most popular open source mapping server and is used by thousands of web mapping sites today. Already in December 1997, HTML 4.0 was published, the first version that allowed styling with CSS and absolute positioning using pixels. Scripting at this time was still highly proprietary. Web authors had to write different script code for the major web browsers: Netscape Navigator and Microsoft Internet Explorer. This period is also well-known as the browser wars, a time in which browser vendors introduced many proprietary features in the HTML language to set themselves apart from the competition, also a time where users and web developers suffered a lot from the incompatibilities between the different web browsers.

In the middle of 1997, UMN Map Server 1.0 was released. It grew into the most popular open source mapping server and is used by thousands of web mapping sites today. Already in December 1997, HTML 4.0 was published, the first version that allowed styling with CSS and absolute positioning using pixels. Scripting at this time was still highly proprietary. Web authors had to write different script code for the major web browsers: Netscape Navigator and Microsoft Internet Explorer. This period is also well-known as the browser wars, a time in which browser vendors introduced many proprietary features in the HTML language to set themselves apart from the competition, also a time where users and web developers suffered a lot from the incompatibilities between the different web browsers.

In June 1998, a joint effort from USGS, Microsoft and HP started the popular US Terra Server project, an OGC compatible WMS server, serving aerial images and USGS topo maps. Shortly after, ESRI entered the web map server market with its Map Objects Internet Map Server, a project that was later replaced by the ArcIMS server. In September 2001, W3C released the SVG 1.0 specification, a XML based mark-up language for the integration of vector graphics, raster graphics and text, also supporting animation, multimedia, scripting and interactivity, and internationalization. SVG opened many opportunities for interactive web maps but it took a while until SVG was properly supported in web browsers. The Tirol Atlas, released in 2001 shortly after SVG 1.0 becoming a recommendation, is the first major online atlas extensively using SVG technology. It is still one of the few atlases offering superior interactivity.

In January 2003, the SVG mobile profiles SVG tiny and SVG basic were released, enabling location based services and mobile maps. In the same year, NASA released World Wind, one of the first interactive virtual earth applications. This software requires a special virtual globe software and does not run in web browsers, but it offers open, OGC compatible interfaces and data is loaded across the internet. In early 2005, Google released Google Maps a web mapping application developed on top of Dynamic HTML, ECMA Script and XML Http Requests (also known as Ajax). The fact that Google opened its API for third party developers for free reuse made it quickly highly popular, triggering thousands of Google Maps derived web mapping applications. Technically, Google Maps is based on quad-tree raster tiles of aerial images and road maps and a geospatial search engine. Later, in the same year, Google released Google Earth, a virtual earth application bought from Keyhole. KML, a XML based mark-up language, allowed users to add their own 3D geodata content. In November 2005, Mozilla released Firefox 1.5, the first version with native SVG support, followed in 2006 by Opera 9, which currently offers the best native SVG support. Apple and KDE will release native SVG support in Safari and Konqueror during 2007. In August 2006, SVG Tiny 1.2 went to W3C candidate recommendation, with improved multimedia support and better support for building rich client internet applications.

Scientific Fundamentals

The advent of web mapping can be regarded as a major new trend in cartography. Previously, cartography was restricted to a few companies, institutes and mapping agencies, requiring expensive and complex hardware and software as well as skilled cartographers and geomatics engineers. With web mapping, freely available mapping technologies and geodata potentially allow every skilled person to produce web maps, with expensive geodata and technical complexity (data harmonization, missing standards) being two of the remaining barriers preventing web mapping from fully going mainstream. The cheap and easy transfer of geodata across the internet allows the integration of distributed data sources, opening opportunities that go beyond the possibilities of disjoint data storage. Everyone with minimal knowhow and infrastructure can become a geo data provider. These facts can be regarded both as an advantage and a disadvantage. While it allows everyone to produce maps and considerably enlarges the audience, it also puts geodata in the hands of untrained people who potentially violate cartographic and geographic principles and introduce flaws during the preparation, analysis and presentation of geographic and cartographic data.

Advantages of Web Maps

Web maps can easily deliver up to date information. If maps are generated automatically from databases, they can display information in almost real-time. They don't need to be printed, mastered and distributed. Imagine a map displaying election results as soon as the election results become available, or a map displaying the traffic situation near real-time by using traffic data collected by sensor networks. Because web maps distribute both logic and data with each request or loading, product updates can happen every time the web user reloads the application. In traditional cartography, a map update caused serious efforts triggering a reprint or premastering as well as a redistribution of the media. With web maps, data and product updates can happen with considerably less efforts and costs, in a much shorter time span and with more dense update intervals. Web server hardware is cheaply available and many open source tools exist for producing web maps.

If web maps are implemented based on open standards, the underlying operating system and browser do not matter. If properly implemented, web maps work across browsers and operating systems. It is also easy to integrate multimedia in and with web maps. Current web browsers support the playback of video, audio and animation. Web maps also support hyperlinking and act as an information hub. By using vector geometries or sensitive areas in a web map, any portion of the map can link to other web pages or web services. As an example, a city map can link to the corresponding timetables of the public transport system at every bus or train station. Web maps can combine distributed data sources. Using open standards and documented APIs one can integrate different data sources, if the projection system, map scale and data quality match. The use of centralized data sources removes the burden for individual organizations to maintain copies of the same data sets. The down side is that one has to rely and trust the external data sources.

By using user profiles, personal filters and personal styling and symbolization, users can personalize, configure and design their own maps. Accessibility issues can be treated in the same way. If users can store their favorite colors, symbols and patterns they can avoid color combinations and map symbolizations they can't easily distinguish, e.g. due to color blindness. Web maps also enable collaborative mapping. Similar to the Wikipedia project, web mapping technologies, such as DHTML/Ajax, SVG, Java, Adobe Flash, etc. enable distributed data acquisition and collaborative efforts. Examples for such projects are the Open Street Map project or the Google Earth community. As with other open projects, quality assurance is essential.

Problems with Web Maps

The realibility of the internet and web server infrastructure is not yet good enough. Especially if a web map relies on external, distributed data sources, the original author cannot guarantee the availability of the information. They usually require a relatively high bandwidth. Despite the increasing availability of free and commercial tools to create web mapping and web GIS applications, web maps are still complex to develop. Many technologies, modules, services and data sources have to be mastered and integrated. Compared to the development of standalone applications with integrated development tools, the development and debugging environments of a conglomerate of different web technologies is still awkward and uncomfortable.

Geodata is often expensive. Unlike in the USA, where geodata collected by governmental institutions is usually available for free or cheap, geodata is usually very expensive in Europe or other parts of the world. This is a serious barrier for many low budget web mapping projects in Europe and decreases quality of web maps due to limited access to high quality geodata. There are also open copyright and privacy issues. Many people are still reluctant to publish geodata, especially in the light that geodata is expensive in some parts of the world. They fear copyright infringements of other people using their data without proper requests for permission. Digital rights managements does not work reliably and is usually cracked in a short time by hackers. With detailed information available and the combination of distributed data sources, it is possible to find out and combine a lot of private and personal information of individual persons. High resolution aerial and satellite images of private properties and estates are now easily accessible throughout the world to anyone.

Finally, as with any other screen based map, there is the problem of limited screen space. This is in particular a problem for mobile web maps and location based services where maps have to be displayed on very small screens with resolutions as low as 100100 pixels. Hopefully, technological advances will help to overcome these limitations.

Types and Properties of Web Maps

One of the first classifications of electronic maps and atlases was carried out by Siekierska and Taylor. They distinguished view-only atlases, atlases that generate maps on demand and analytical atlases based on GIS capabilities. Ormeling used the terms view-only atlases, interactive atlases and analytical atlases for his classification of electronic atlases. A first classification of web maps was made by Kraak. He distinguished between static and dynamic web maps and further differentiated between interactive and view only web maps. However, today in the light of an increased number of different web map types, this classification needs some revision. There are additional possibilities regarding distributed data sources, collaborative maps, personalized maps, and many more. It is impossible to create a complete classification covering all types of web maps. However, various properties of a web map can be defined. Figure lists potential properties of web maps with each property as a value-pair. Every product can be allocated to either property of the value pair or in between on a scale between 0 and 1.

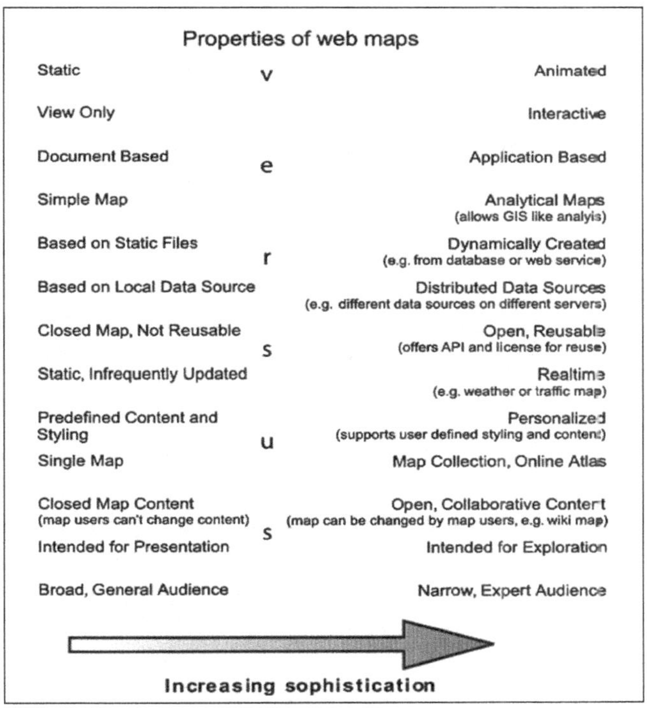

Potential properties of web maps.

This list of properties is not complete and more properties can be added as web maps gain capabilities.

Following is a list of web map types with a short description of their properties and characteristics. Obviously, many maps can be allocated to more than one web map type.

Static web maps are view only with no animation and interactivity. They are only created once, often manually and infrequently updated. Typical graphics formats for static web maps are png, jpeg or gif for raster files, svg, pdf or swf for vector files. Often, these maps are scanned paper maps and had not been designed as screen maps. Paper maps have a much higher resolution and information density and might be illegible when displayed on screens at the wrong resolution.

Dynamically created web maps are created on demand each time the user reloads the webpages, often from dynamic data sources, such as databases. The webserver generates the map using a web map server or a self-written software.

Distributed web maps are created from distributed data sources. The WMS protocol offers a standardized method to access maps on other servers. WMS servers can collect these different sources, reproject the map layers, if necessary, and send them back as a combined image containing all requested map layers. One server may offer a topographic base map, while other servers may offer thematic layers.

Animated web maps show changes in the map over time by animating one of the graphical or temporal variables. Various data and multimedia formats and technologies allow the display of animated web maps: SVG, Adobe Flash, Java, QuickTime, etc., also with varying degrees of interaction. Examples for animated web maps are weather maps, maps displaying dynamic natural or other phenomena (such as water currents, wind patterns, traffic flow, trade flow, communication patterns, etc.).

Real-time web maps show the situation of a phenomena in close to real-time (only a few seconds or minutes delay). Data is collected from sensors, sent across the internet to a central server, and the maps are generated or updated at regular intervals or immediately on demand. Examples are weather maps, traffic maps or vehicle monitoring systems.

Personalized web maps allow the map user to apply his own data filtering, selective content and the application of personal styling and map symbolization. The OGC consortium provides the SLD standard (Styled Layer Description) that may be sent to a WMS server for the application of individual styles. This implies that the content and data structure of the remote WMS server is properly documented.

Open, reusable web maps are usually more complex web mapping systems that offer APIs for reuse in other people's web pages and products. An example for such a system is Google Maps with the Google Maps API. Ideally, such APIs would be compatible with the standards promoted by the Open Geospatial and W3C Consortium, but unfortunately, in reality they are often proprietary.

Interactive web maps help to compensate for the disadvantages of screen and web maps (limited screen space, bad resolution, limited color ranges, etc.). Interactivity helps to explore maps, change map parameters, navigate and interact with the map, reveal additional information, link to other resources, and much more. Technically, it is achieved through the combination of events, scripting and DOM manipulations.

Analytic web maps offer GIS analysis, either with geodata provided, or with geodata uploaded by the map user. As already mentioned, the borderline between analytic web maps and web GIS is blurry. Often, parts of the analysis is carried out by a server side GIS and the client displays the result of the analysis. As web clients gain more and more capabilities, this task sharing may gradually shift.

Collaborative web maps are still new, immature and complex to implement, but show a lot of potential. The idea is that, like in the Wikipedia project, various, distributed, people collaborate to create and improve maps on the web. Technically, an application allowing simultaneous editing across the web would have to ensure that geometric features being edited by one person are locked, so that they can't be edited by other persons at the same time. Also, a minimal quality check should be made, before data goes public. Some projects working on collaborative maps are Open Street Map, Google Earth and WikiMapia.

Web Mapping Architectures

The potential number of technologies to implement web mapping projects is almost infinite. Any programming environment, programming language and server side framework can be used to implement web mapping projects. In any case, both server and client side technologies have to be used. Following is a list of potential and popular server and client side technologies utilized for web mapping. Figure shows one potential web mapping architecture that was used for a web map on Slope Stability on Nisyros Island (Greece). It includes the most important components of a web mapping architecture, but obviously some modules could be replaced by other software and some modules aren't needed for certain types of web maps.

Server Side Technologies

The webserver is responsible for handling http requests by web browsers and other user agents. In the simplest case they serve static files, such as HTML pages or static image files. Web servers also handle authentication, content negotiation, server side includes, URL rewriting and forward requests to dynamic resources, such as CGI applications or server side scripting languages. The functionality of a webserver can usually be enhanced using modules or extensions. The most popular web server is Apache, followed by Microsoft Internet Information Server, Netscape and others.

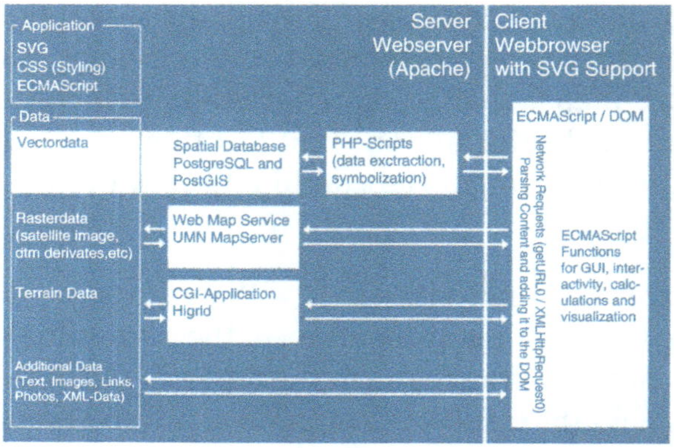

Example of a web mapping architecture.

CGI (common gateway interface) applications are executable running on the webserver under the environment and user permissions of the webserver user. They may be written in any programming language (compiled) or scripting language (e.g. perl). A CGI application implements the common gateway interface protocol, processes the information sent by the client, does whatever the application should do and sends the result back in a web-readable form to the client. As an example, a web browser may send a request to a CGI application for getting a map with a certain map extent, styling and map layer combination. The result is an image format, e.g. jpeg, png or SVG. For performance enhancements one can also install CGI applications as Fast CGI. This loads the application after the web server is started and keeps the application in memory, eliminating the need to spawn a separate process each time a request is being made. Alternatively, one can use scripting languages built into the webserver as a module, such as PHP, Perl, Python, ASP, Ruby, etc. If built into the web server as a module, the scripting engine is already loaded and doesn't have to be loaded each time a request is being made.

Web application servers are middleware which connect various software components with the web server and a programming language. As an example, a web application server can enable the communication between the API of a GIS and the webserver, a spatial database or other proprietary applications. Typical web application servers are written in Java, C, C++, C# or other scripting languages. Web application servers are also useful when developing complex real-time web mapping applications or Web GIS.

Spatial databases are usually object relational databases enhanced with geographic data types, methods and properties. They are necessary whenever a web mapping application has to deal with frequently changing dynamic data or with huge amounts of geographic data. Spatial databases allow spatial queries, sub selects, reprojections, geometry manipulations and offer various import and export formats. A popular example for an open source spatial database is Post GIS. MySQL also implements some spatial features, although not as mature as Post GIS. Commercial alternatives are Oracle Spatial or spatial extensions of Microsoft SQL Server and IBM DB2. The OGC Simple

Features for SQL Specification is a standard geometry data model and operator set for spatial databases. Most spatial databases implement this OGC standard.

WMS server are specialized web mapping servers implemented as a CGI application, Java Servlet or other web application server. They either work as a standalone web server or in collaboration with existing web servers or web application servers (the general case). WMS Server can generate maps on request, using parameters, such as map layer order, styling/symbolization, map extent, data format, projection, etc. The OGC Consortium defined the WMS standard including the map requests and allowed return data formats. Typical image formats for the map result are png, jpeg, gif or SVG. An open source WMS Server is the UMN Map server. Commercial alternatives exist from most commercial GIS vendors, such as ESRI ArcIMS, Intergraph Geomedia Web Map and others.

Client Side Technologies

In the simplest setup, only a web browser is required. All modern web browsers support the display of HTML and raster images (jpeg, png and gif format). Some solutions require additional plugins. ECMA Script is the standardized version of JavaScript. It is necessary to implement client side interaction, refactoring of the DOM of a webpage and for doing network requests. ECMA Script is currently part of any modern web browser. Various events are necessary to implement interactive client side maps. Events can trigger script execution or SMIL operations. One distinguishes between mouse events, keyboard events, state events, mutation events, SMIL animation events, UI events and SVG specific events. Network requests are necessary to load additional data and content into a web page. Most modern browsers provide the XMLHttp Request object which allows get and post http requests and provides some feedback on the data loading state. The data received can be processed by ECMA Script and can be included into the current DOM tree of the web page or web map. SVG user agents alternatively provide the get URL and post URL methods for network requests. These network requests are also known under the term Ajax.

The Document Object Model provides a language independent API for the manipulation of the document tree of the webpage. It exposes properties of the individual nodes of the document tree, allows to insert new nodes, delete nodes, reorder nodes and change existing nodes. DOM support is included in any modern web browser. DOM support together with scripting is also known as DHTML or Dynamic HTML. Google Maps, Microsoft Local and many other web mapping sites use a combination of DHTML, Ajax, SVG and VML. SVG is the abbreviation of Scalable Vector Graphics and integrates vector graphics, raster graphics and text. SVG also supports animation, internationalization, interactivity, scripting and XML based extension mechanisms. SVG is a huge step forward when it comes to delivering high quality, interactive maps. At the time of writing, SVG is natively supported in Mozilla Firefox (version 1.5 or later), Opera (version 9 or later) and the developer version of Safari Webkit. Internet Explorer users still need the Adobe SVG viewer plugin provided by Adobe. Neumann and Winter discusses the use of SVG for interactive web mapping.

Since 1995, Java applets can be used in web browsers. Some browsers still provide old versions of the Java virtual machine. An alternative is the use of the Sun Java Plugin. Java is a full featured programming language that can be used to create very sophisticated and interactive web maps. The Java2D and Java3D libraries provide 2d and 3d vector graphics support. The creation of Java based web maps requires a lot of programming know how. Herzog discusses the use of Java applets for the presentation of interactive choroplethe maps and cartograms.

Web browser plugins are a way to extend the capabilities of a web browser. While this works fine in theory, browser plugins create a lot of problems, because plugins often cannot interact properly with the rest of the browser and cannot be blended easily with other contents of a web page. A popular plugin is Adobe Acrobat for viewing PDF files. PDF files may display map layers (they can be interactively turned on and off) and contain interactivity, links and scripting. Another popular plugin is Adobe Flash (former Macromedia), providing vector graphics, animation and multimedia support. It allows the distribution of sophisticated interactive maps, as with Java and SVG. It also features a programming language (Action Script) which is similar to ECMA Script and supports Audio and Video. A gallery of highly interactive Adobe Flash based maps can be seen at Mauvière and Coulomb. Apple QuickTime adds support for additional image formats, video, audio and QuickTime VR (Panorama Images). The Adobe SVG viewer provides SVG 1.0 support for web browsers with no native SVG support. The Sun Java plugin provides support for newer and advanced Java Features.

Key Applications

Address Matching and Routing

A classic use case of web maps are the display of the results of address searches and route finding operations. Typically, the searches and route finding algorithms are executed on the web server and the resulting map sent back to the client, with the search result highlighted. The results of a route finding operation may be animated and interactivity can be used for the display of additional information (such as points of interest, accompanying text at junctions, etc.).

Real Time Maps

Real time web maps fully exploit the strengths of web maps. Sensors send their data to a central server repository which is used to generate real time (or near real time) maps on demand. Examples are weather maps, traffic maps, maps for fleet management, etc. Once a map is loaded it can update the dynamic information on demand, without having to reload the application or base map. This is technically implemented using network requests (XMLHttp Requests or sockets).

Location based Services

Location based services (LBS) need maps to display points of interest. Mobile web maps are displayed by relying on mobile browsers, SVG, Flash or Java based user agents. Mobile web maps need to be specially designed, taking into account the small screen space and low resolution of mobile displays. Most of the time, maps for location based services are generated automatically from spatial databases.

Urban and Regional Planning

Web maps can support the urban and regional planning process by informing the concerned public and decision makers through a visualization of planned activities and measures. Depending on the type of the web map, web maps can be used for presentation purposes or also enable a feedback process by letting the public provide feedback using interactive maps. As an example, a web map could present different scenarios and let the public vote and help decide which action should be taken. A more advanced version would allow the public to directly create and manipulate features in a web map to create their own, preferred scenario. Animated maps are especially useful when presenting simulations or visualizing measures that have different effects over time.

Online Atlases

Atlas projects often went through a renaissance when they made a transition to a web based project. In the past, atlas projects often suffered from expensive map production, small circulation and limited audience. Updates were expensive to produce and took a long time until they hit the public. Many atlas projects, after moving to the web, can now reach a wider audience, be produced cheaper, provide a larger number of maps and map types and integrate with and benefit from other web resources. The first atlas which went online is the Atlas of Canada, followed by the US National Atlas and commercial offerings, such as Microsoft Encarta Online. The Tirol Atlas is the first online atlas making substantial use of SVG and open source web mapping technology. Some atlases even ceased their printed editions after going online, some-times offering printing on demand features from the online edition. Some atlases (primarily from North America) also offer raw data downloads of the underlying geospatial data sources.

E-Learning

Web maps can be included in e-learning systems. They can illustrate geospatial phenomena, provide a training environment of spatial planning or operations and may even be used as test environments, testing the spatial knowledge and abilities of students. Interactivity, multimedia and animation may support the learning process.

Collaborative Mapping

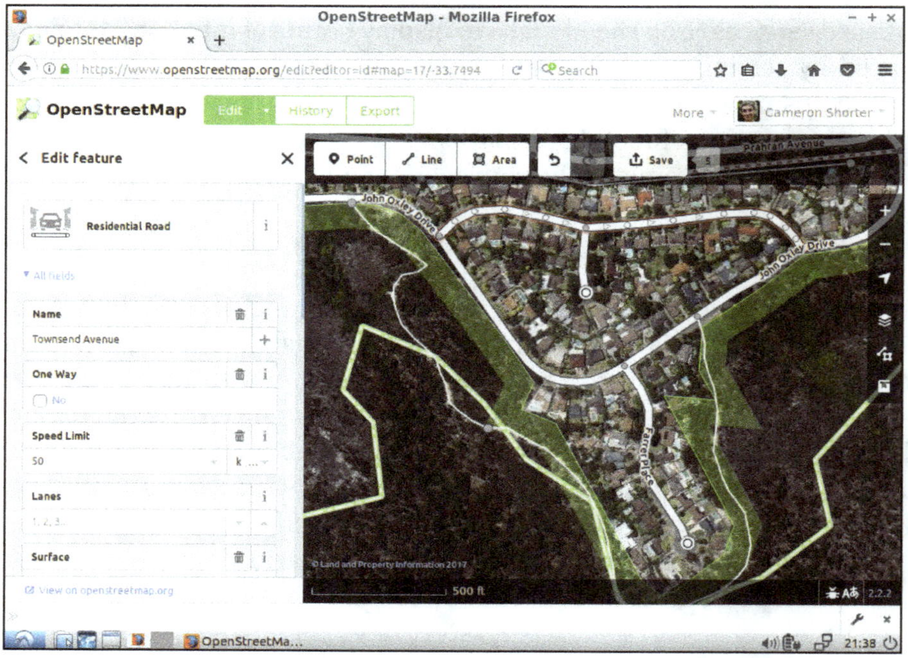

Mapping OpenStreetMap with the iD editor, done within the browser.

Collaborative mapping is the aggregation of Web mapping and user-generated content, from a group of individuals or entities, and can take several distinct forms. With the growth of technology for storing and sharing maps, collaborative maps have become competitors to commercial services, in the case of OpenStreetMap, or components of them, as in Google Map Maker and Yandex.Map editor.

Volunteers collect geographic information and the citizens/individuals can be regarded as sensors within a geographical environment that create, assemble, and disseminate geographic data provided voluntarily by the individuals.Collaborative mapping is a special case of the larger phenomenon known as crowd sourcing, that allows citizens to be part of collaborative approach to accomplish a goal. The goals in collaborative mapping have a geographical aspect, e.g. having a more active role in urban planning. Especially when data, information, knowledge is distributed in a population and an aggregation of data is not available, then collaborative mapping can bring a benefit for the citizens or activities in a community with an e-Planing Platform. Extensions of critical and participatory approaches to geographic information systems combines software tools with a joint activities to accomplish a community goal. Additionally, the aggregated data can be used for a Location-based service like available public transport options at the geolocation where a mobile device is currently used (GPS-sensor). The relevance for the user at a specific geolocation cannot be represented with logic value in general (relevant=true/false). The relevance can be represented with Fuzzy-Logic or a Fuzzy architectural spatial analysis.

Types

Collaborative mapping applications vary depending on which feature the collaborative edition takes place: on the map itself (shared surface), or on overlays to the map. A very simple collaborative mapping application would just plot users' locations (social mapping or geosocial networking) or locations (Placeopedia). Collaborative implies the possibility of edition by several distinct individuals so the term would tend to exclude applications where the maps are not meant for the general user to modify.

In this kind of application, the map itself is created collaboratively by sharing a common surface. For example, both OpenStreetMap and WikiMapia allow for the creation of single ‹points of interest', as well as linear features and areas. Collaborative mapping and specifically surface sharing faces the same problems as revision control, namely concurrent access issues and versioning. In addition to these problems, collaborative maps must deal with the difficult issue of cluttering, due to the geometric constraints inherent in the media. One approach to this problem is using overlays, allowing to suitable use in consumer services. Despite these issues, collaborative mapping platforms such as OpenStreetMap can be considered as being as trustworthy as professionally produced maps.

Overlays group together items on a map, allowing the user of the map to toggle the overlay's visibility and thus all items contained in the overlay. The application uses map tiles from a third-party (for example one of the mapping APIs) and adds its own collaboratively edited overlays to them, sometimes in a wiki fashion. If each user's revisions are contained in an overlay, the issue of revision control and cluttering can be mitigated. One example of this is the accessibility platform Accessadvisr, which utilises collaborative mapping to inform persons of accessibility issues, which is perceived to be as reliable and trustworthy as professional information.

Other overlays-based collaborative mapping tools follow a different approach and focus on user centered content creation and experience. There users enrich maps with their own points of interest and build kind of travel books for themselves. At the same time users can explore overlays of other users as collaborative extension.

Humanitarian Collaborative Mapping

Humanitarian OpenStreetMap Team, based on OpenStreetMap, provides collaborative mapping support for humanitarian objectives, e.g. collaborative transportation map, epidemiological mapping for Malaria, earthquake response, or typhoon response.

Collaborative 3D Mapping

In robot navigation, 3-dimensional maps can be reconstructed collaboratively using simultaneous localization and mapping.

Private Local Collaboration using Maps

Some mapping companies offer an online mapping tool that allows private collaboration between users when mapping sensitive data on digital maps, e.g.:

- Google Maps,
- Wegovnow: A map based platform to engage the local civic society – local collaboration & publishing with maps.
- Canvis.app: A platform that allows you to easily generate, customize, and share a collaborative mapping campaign. Suitable for large scale crowdsourcing projects.

Quality Assurance

If citizens or a community collects data, information then concerns come up about data quality, and specifically about its credibility. The same aspects of quality assurance are relevant for collaborative mapping and the possibility of vandalism.

Data Collection Tools

Collaborative mapping is not restricted to the application of mobile devices but if data is captured with a mobile device the satellite navigation (like GPS is helpful to assign the current geolocation to the collected data at the geolocation. Open Source tools like Open Data Kit are used to collect the mapping data (e.g. about health care facilities or humanitarian operations) with a survey that could automatically insert the geolocation into the survey data that could include visual information (e.g. images, videos) and audio samples collected at the current geolocation. An image can be used e.g. as additional information of damage assessment after an earth quake.

Restricted Visibility of Alterations

These sites provide general base map information and allow users to create their own content by marking locations where various events occurred or certain features exist, but aren't already shown on the base map. Some examples include 311-style request systems and 3D spatial technology.

Public Alterations and Quality Assured Versions

The openness for changes to the community is possible for all individuals and the community is validating changes by putting regions and location at their personal watchlist. Any changes in the joint repository of the mapping process are captured by a version control system- Reverting changes is possible and specific quality assured versions of specific areas can be marked as reference map for a specific area. Quality assurance can be implemented on different scales:

- Version of complete map,
- Version of selected regions/area,
- Version of mapping attributes a Point of Interest (e.g. hospital marked as "under construction" is providing health care services).

Blockchain can be used as integrity check of alterations or digital signature can be used to mark a certain version as "quality assured" by the institution that signed a map as digital file or digital content.

WEB MAPS

"Web map" often implies a map that is not simply on the web, but rather one that is powered by the web. A digital map is on a computer, but may not be accessible by internet, and is relatively static if it is. A web map depends on the internet. It is usually interactive and not always self-contained.

Powered by the Web

Web maps, in contrast to a simple static map image that resides on the web, rely on the web in order to exist and function. This means one or more of a few things:

- Web technology: web maps are built with the technology of the web, such as HTML, JavaScript, and CSS. (Most web maps involve at least those three in particular.) A web map runs on technology that can be consumed by standard web browsers, including, ideally, mobile web browsers.
- Server-side communication: a web map might rely on the internet to provide its data (say, stored in an online database) or perform calculations. This means that web maps can contain vastly more information than can possibly be presented to you at any single moment. It's quite a powerful contrast to the paper maps of old—imagine a map with such massive amounts of information as Google Maps, and trying to fold that up into pocket!
- Drawing from outside sources: on the web it's not always necessary to collect all data sources in one place before mapping them—the map itself is that collection. Many data providers have APIs into which a map can tap directly every time a user views the map, eliminating the need to download all that data beforehand.

- Real-time data: being connected to the internet, many web maps make use of real-time, live data feeds so that they are constantly up to date on whatever they're showing. This, of course, is another thing that was never possible with traditional maps, which always necessarily represented a snapshot in time.

Anatomy of a Web Map

In many contexts "web map" has an even narrower definition. It includes some of the maps you probably use almost every day, such as Google Maps.

Slippy Maps

The typical form of web map is sometimes called a "slippy map." This is a type of map you're used to, like Google's: you can grab and pan the map, and zoom in or out. Google was a major pioneer of this type of map with the introduction of their map service in 2005. It represented a huge leap from the previous standard web map, where panning (in fixed amounts) or zooming essentially required reloading the entire map image with a new extent.

Tiles

Slippy maps accomplish their slick usability with tiling schemes. In the old days, a map view was a single image. You had to load the whole thing before you could see it. Now, the map you see comprises many smaller tiles that load invididually. What this means is you only load the part of the map you need. For example if you want to look slightly east, you only need to load a new column of tiles on the right side instead of reloading the entire map. Tiling makes map viewing fast and data much more manageable.

Tiles come in two flavors: raster tiles are pre-rendered images of map data, while vector tiles are small sections of the data itself, which is then rendered in the browser. Raster tiles are the original standard and are still very common, but vector tiles have moved to the forefront as technology has made live-rendering of map data feasible. Vector tiles are efficient and fast for a variety of reasons.

Tiled maps use a standard tiling scheme which is worth understanding. Tiles are idenfied by three numbers, typically in this order:

- Zoom level: The scale of the map for this tile, relative to an initial scale at which the whole world fits in one tile.
- An X coordinate: The horizontal position of this tile, within the grid of tiles for the world map at this zoom level. Usually, this number increases from left to right, with 0 being the left edge of the map, or 180° West longitude.

- A Y coordinate: The vertical position of this tile, within the grid of tiles for the world map at this zoom level. Usually, this number counts up from top to bottom to right, with 0 being the top edge of the map, around 85° North latitude. (In the standard Mercator map projection of web maps, 85° latitude northern and southern limits allows everything to fit in a square.)

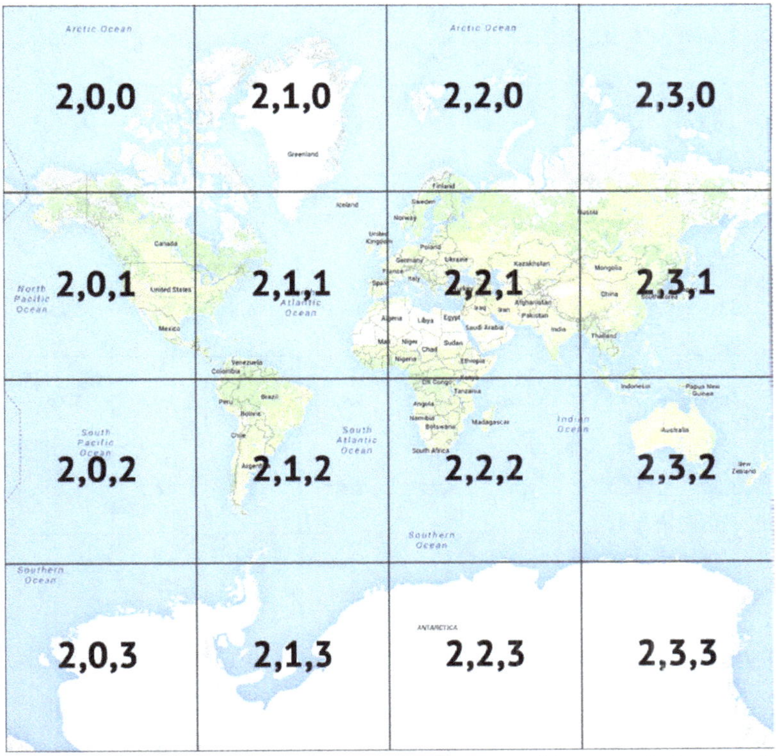

Tiles are typically numbered with [z, x, y] coordinates. This example is zoom level 2.

Each zoom level contains twice as many rows and columns as the previous level. With zoom level zero being a single tile for the whole world, this means the number of tiles for any zoom level z is:

$2^z \times 2^z$, or 4^z

Notice the relationship of tile numbers in zoom level 3 to those in zoom level 2. Consider the top left corner of each tile in zoom level 2. When we increase the zoom level by one, the x and y for a tile with that same corner position on the map are multiplied by two. For example Norway is in tile [2,2,1] at zoom level 2, and [3,4,2] at zoom level 3.

Although the numbering pattern is different in a couple of systems, tile positions and extents are nearly universal and are well understood by code libraries such as Leaflet. Many are built to understand URL patterns for tile locations (tiles are stored online one way or another).

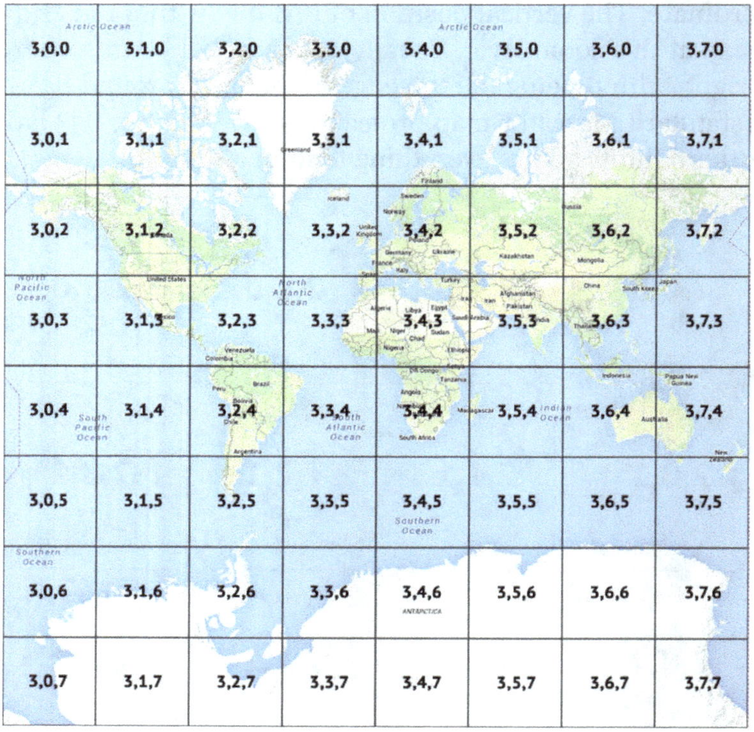

Map Projections on the Web

Standard slippy maps use the Web Mercator projection, much to the chagrin of many cartographers. There are a couple good reasons for this though:

- The map is conformal (preserving shape) and represents all constant bearings as straight lines. What this boils down to is that you can zoom into the map anywhere in the world, and all shapes and directions will look more or less correct. This isn't true of other map projections, where, for example, Quito might look right but a high-latitude place like Anchorage would look terribly squished.

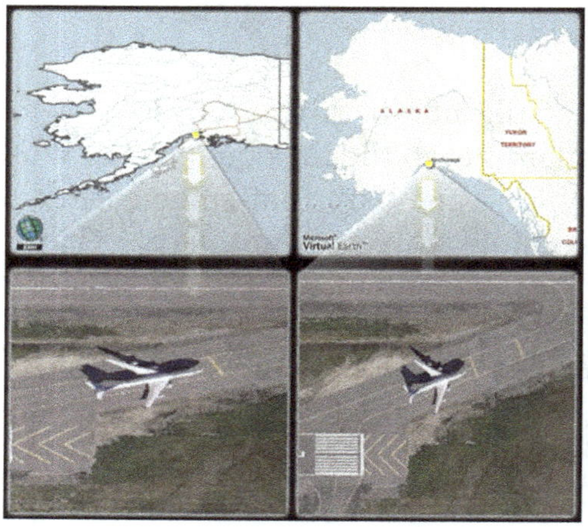

- Almost the entire world fits neatly in a square. This makes the tiling scheme described above work very well.

However, it comes with a few pitfalls:

- The Mercator projection is wildly inappropriate for some types of thematic maps. (To say nothing of the generally distorted version of the world it presents.) Choropleth maps and others that depend on correct area proportions should not be used with Mercator maps at small (global-to-continental) scales, as area distortions can be huge. Distortions are minimal at local levels, though.
- As implied above, scale is not constant around the world. Be suspicious if someone presents you with Google Maps screenshots of two different places "at the same scale." Zoom level is not the same as scale. At high latitudes, where areas are enlarged, a tile represents less area than near the equator.

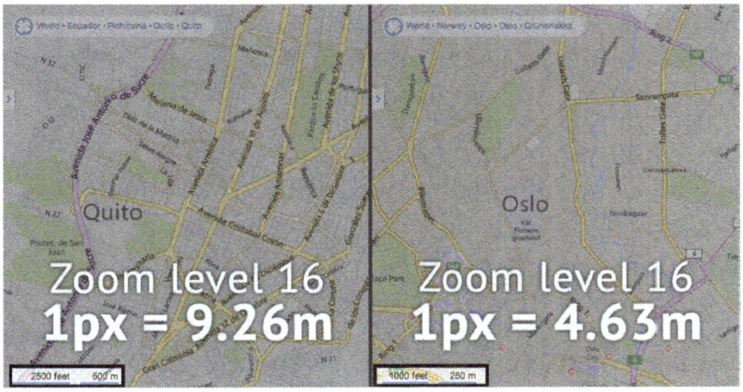

- Although most of the world fits in a square, not all of it does. In fact, it's impossible to represent the north and south pole at all—they stretch infinitely up and down! While the typical map, which cuts off at about 85 degrees north and south, doesn't miss much, it does miss something, notably much of Antarctica.

Yahoo! Maps

Yahoo! Maps was a free online mapping portal provided by Yahoo! Functionality included local weather powered by The Weather Channel, printing maps, and local reviews powered by Yelp. It shut down in 2015.

Map Data

The street network and other vector data Yahoo! Maps used later on was from HERE, and includes a number of public data sources. Detailed street network data is currently available for the United States, Canada, Puerto Rico, the Virgin Islands, and most European countries. Country borders, cities, and water bodies are mapped for the rest of the world.

Low-resolution satellite imagery is available worldwide. 1–2 meter resolution is available for most of the contiguous United States and select cities worldwide.

Functions

The main Yahoo! Maps site offers street maps and driving directions for the United States and Canada. It has the following notable features:

- Address Book: Registered Yahoo! users can store a list of commonly used street addresses, making it unnecessary to type them in again. A recently entered address can be quickly recalled by selecting one from a drop-down list.
- Live Traffic: Traffic incident markers and current highway conditions can be viewed on the map.
- Find On The Map: A local search by business name or category can be typed into the "Find On The Map" box to locate it in the current map view. A list of clickable point of interest categories is also available. The results can be further refined by user rating, or related category.
- Driving Directions: Driving directions can be displayed on a map or in printable form, with optional turn-by-turn maps, or as simple text. Links to driving directions can be e-mailed, and text directions sent to mobile phones. Multi-point driving directions: Multiple addresses can be entered and manually reordered for complex driving directions.
- Draggable maps: The map view can be manipulated by dragging it with the mouse or tapping the arrow keys. Zoom level can be controlled via the mouse scroll wheel, "Page Up"/"Page Down" keys, or the map's zoom bar.
- Widgets: A number of widgets over the map include a navigator widget, map type (map, satellite & hybrid) controller and a zoom level control.
- Satellite Imagery: Labelled (hybrid) and unlabelled satellite imagery is available worldwide.
- Overview map: Collapsible overview map provides context, with draggable grey area controlling the main map view.
- International Coverage: Outside the US and Canada, Yahoo! Maps Beta can recognize city, province, and country names, and provide a small-scale map or satellite views.
- Right click to set waypoint: an origin, destination, or midpoint can be set by right-clicking on the desired location on the map.
- Draggable markers: Any marker can be dragged to the 'Get Map' text entry area to add that location to a route.

Developer API

Until December 2011, developers could embed Yahoo! Maps into their own web pages (to create a mashup) through the Yahoo! Maps Developer APIs. The Yahoo! Maps APIs came in three basic flavors:

- The Flash APIs, which used the Adobe Flash platform. Three variations, allowing the developer to write in Java Script, Action Script, or Adobe Flex 1.5, were available.
- The Ajax API, for interactive maps that use capabilities inherent in web browsers, without using the Flash plug-in. Ajax applications are written in JavaScript.
- The "Simple" API, basically an XML data format, an extension of Geo RSS, for displaying point of interest data on top of Yahoo!'s main map site. The Flash and Ajax APIs also supported display of Geo RSS formatted data.

Yahoo! offered a number of low-level APIs to support maps, for geocoding, getting a map image, searching for a local business, or retrieving traffic information. Some other Yahoo! services, such as Flickr and Upcoming.org, have their content available through web services, with interesting potential for mashups.

Bing Maps

Bing Maps (previously Live Search Maps, Windows Live Maps, Windows Live Local, and MSN Virtual Earth) is a web mapping service provided as a part of Microsoft's Bing suite of search engines and powered by the Bing Maps for Enterprise framework.

Windows Live Local Beta, showing aerial imagery

Bing Maps was originally launched as MSN Virtual Earth, which was released for beta testing on July 24, 2005. It was a continuation of previous Microsoft technologies such as Microsoft MapPoint and TerraServer. Its original stand out feature was the aerial imagery. The original version lacked many of its distinguishing features, including birds' eye view and 3D maps, and the Collections functionality was limited to a single "Scratchpad" of points of interest.

In December 2005, Virtual Earth was replaced by Windows Live Local, featuring improvements, technologies from Pictometry International, and integrated with the Local Search index on Windows Live Search. On November 6, 2006, Microsoft added the ability to view the maps in 3D using a .NET managed control and managed interfaces to Direct3D. Microsoft subsequently referred to this product officially as "*Live Search Maps*", integrating it as part of its Live Search services.

On June 3, 2009, Microsoft officially rebranded Live Search Maps as *Bing Maps*, and the Virtual Earth platform as *Bing Maps for Enterprise*.

In 2010, Microsoft added an OpenStreetMap layer to Bing Maps. From 2012, Nokia (formerly Navteq) powered many aspects of Bing Maps as an extension to its Windows Phone 7 partnership with Microsoft, including mapping data, geocoding, traffic data and navigation. As of 2018, Here (formerly Nokia) continues to provide street data for Bing Maps.

Updates

- v1 (Beagle) (July 2005),
- v2 (Calypso) (December 2005) - "Bird's-eye imagery" released,
- v2.5 (February 2006),
- v3 (Discovery) (May 2006) - Real time traffic, collections, new API,
- v4 (Endeavour) (September 2006) - People search, drawing on maps, new imagery,
- v5 (Spaceland) (November 2006) - 3D viewer, building models in 15 cities,
- Data update (December 2006) - New 3D models and high-resolution imagery for 6 new areas,
- Data update (January 2007) - Over 100 European cities with bird's-eye coverage added,
- Data update (29 March 2007) - 3.8TB of bird's-eye imagery, orthophotos and 3D models of 5 British cities,
- v5.5 (Falcon) (3 April 2007) VE 3D plugin for Firefox, GeoRSS support, area calculations,

- v6 (Gemini) (15 October 2007) - New data, party maps, traffic based routing, v6 MapControl, Bird's Eye in 3D, etc.,
- v6.1 (GoliatH) (10 April 2008) - Improved quality of 3D models, improved KML support and new export capabilities, street labels on Bird's Eye imagery, Map-Cruncher integration, HD filming capabilities, Clearflow traffic report system,
- v6.2 (Helios) (24 September 2008) - Multi-point driving directions, landmarks in directions, weather, real stars, new data,
- Data Update (29 December 2008) - 48TB of road network data,
- v6.2 (Ikonos) (14 April 2009) - Performance improvements,
- Bing (3 June 2009),
- Bing Maps Silverlight Beta (2 December 2009) - Silverlight, Twitter, Streetside,
- (Oslo) (11 June 2010) - Silverlight improvements,
- (Boston M4) (December 2010) - New map style Venue maps.

Imagery Updates

Bing maps frequently update and expand the geographic areas covered by their imagery, with new updates being released on roughly a monthly basis. Each imagery release typically contains more than 10TB of imagery.

However, the necessary time-lapse before images are updated means that aerial and Bird's-Eye images for a particular location can sometimes be several years out-of-date. This is particularly noticeable in locations that have undergone rapid recent development or experienced other dramatic changes since the imagery was taken, such as areas affected by natural disasters.

Features

Street Maps

Users can browse and search topographically-shaded street maps for many cities worldwide. Maps include certain points of interest built in, such as metro stations, stadiums, hospitals, and other facilities. It is also possible to browse public user-created points of interest. Searches can cover public collections, businesses or types of business, locations, or people. Five street map views are available: Road View, Aerial View, Bird's Eye View, Street Side View, and 3D View.

- Road View: Road view is the default map view and displays vector imagery of roads, buildings, and geography. The data from which the default road map is rendered is licensed from Navteq. In certain parts of the world, road view maps from alternative data providers are also available. For example, when viewing a map of London, the user may see road data from the Collins Bartholomew London Street Map. In all parts of the UK, road data from the Ordnance Survey can also be displayed. A Bing Maps app is available that will display road data from OpenStreetMap.

- Aerial View: Aerial view overlays satellite imagery onto the map and highlights roads and major landmarks for easy identification amongst the satellite images. Since end of November 2010, OpenStreetMap mappers have been able to use imagery of Bing Aerial as a map background.

  At the end of January 2012, both Bing Aerial and Birds Eye View imagery at military bases in Germany became blurred. This was on request of the German government obviously using data of OpenStreetMap.

- Bird's-eye View: Bird's-eye view displays aerial imagery captured from low-flying aircraft. Unlike the top-down aerial view captured by satellite, Bird's-eye images are taken at an oblique 45-degree angle, showing the sides and roofs of buildings giving better depth perception for geography. With Bird's Eye views, many details such as signs, advertisements and pedestrians are clearly visible. Microsoft has occasionally removed Bird's Eye view from areas where it was previously available.

- Streetside: Streetside provides 360-degree imagery of street-level scenes taken from special cameras mounted on moving vehicles. Launched in December 2009 it contains imagery for selected metro areas in the United States as well as selected areas in Vancouver and Whistler, British Columbia associated with the 2010 Winter Olympic Games (example: Richmond Olympic Oval). Selected cities in Europe were also made available in May 2012.

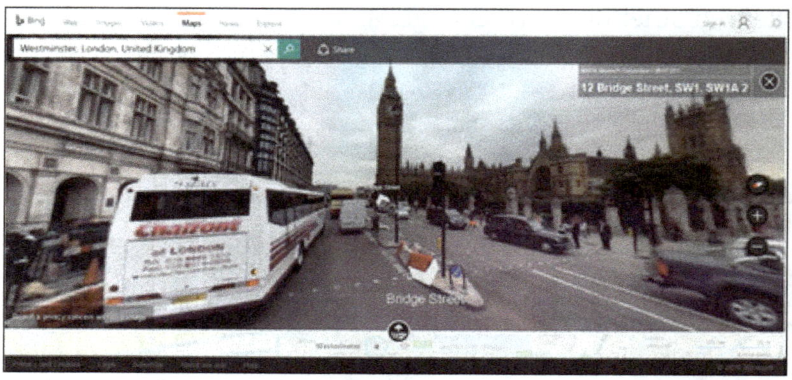

Bing Maps showing Streetside's view near the Palace of Westminster.

Between August and September 2011, German customers were allowed to appeal against integration of their house or flat in Bing Streetside. According to some officials, the number of appeals was significantly lower than with Google Street View. Only 40,000 requests were sent to Microsoft.

Bing Maps Streetside car with cameras on the roof.

For OpenStreetMap editors, display of Streetside tracks and images can be enabled via a map data layer checkbox.

- VenueMaps: Venue maps provides a way of seeing the layout of the venue. Currently, Bing Maps provides maps & level wise layouts of over 5300 venues across the world. The categories are: Airports, Amusement Parks, Buildings, Convention Centers, Hospitals, Malls, Museums, Parks, Racecourses, Racetracks, Resorts, Shopping Centers, Shopping Districts, Stadiums, Universities and Zoos.

- 3D Maps: The 3D maps feature allows users to see the environment (e.g. buildings) in 3D, with the added ability to rotate and tilt the angle in addition to panning and zooming. To attempt to achieve near-photorealism, all 3D buildings are textured using composites of aerial photography. To view the 3D maps, users must install a plugin, then enable the "3D" option on "Bing Maps". In addition to exploring the maps using a mouse and keyboard, it is possible to navigate the 3D environment using an Xbox 360 controller or another game controller in Windows 7, Windows Vista or Windows XP.

  More than 60 cities worldwide could be viewed in 3D, including most of the major cities in the United States and a few cities in Canada, the United Kingdom, and France. Some additional cities have had a select few important landmarks modelled in 3D, such as the Colosseum in Rome. Terrain data is available for the entire world. It is also possible to use a 3D modelling program called 3DVIA Shape for Maps to add one's own models to the 3D map. Since 2014, new 3D imagery has been introduced to a number of new cities.

Driving, Walking and Transit Directions

Users can get directions between two or more locations. In September 2010, Bing Maps added public transit directions (bus, subway, and local rail) to its available direction options. Currently transit directions are only available in 11 cities: Boston, Chicago, Los Angeles, Minneapolis, Newark Metro Area, New York Metro Area, Philadelphia, San Francisco, Seattle, Vancouver BC, and Washington DC. It is also available in other countries such as Spain, Germany, Italy, Austria, Brazil, Mexico and some others.

Map Apps

Bing Map Apps is a collection of 1st and 3rd party applications that add additional functionality and content to Bing Maps. Examples of map apps include a parking finder, a taxi fare calculator, an app that maps out Facebook friends, and an app which lets users explore the day's newspaper front pages from around the world. These apps are only accessible through Bing Maps Silverlight. A source code is available on Microsoft Developer Network to explain integration of Maps in Web Applications.

Traffic Information and Clear Flow

Bing Maps shows users current traffic information for major highways and roads. The feature uses 4 color codes (black, red, yellow, green) to indicate traffic volume, from heaviest traffic to lightest traffic. Microsoft announced in March 2008 that it will release its latest software technology called "ClearFlow". It is a Web-based service for traffic-based driving directions available on Bing.com in 72 cities across the U.S. The tool took five years for Microsoft's Artificial Intelligence team to develop. ClearFlow provides real-time traffic data to help drivers avoid traffic congestion. ClearFlow gives information for alternative routes and supplies traffic conditions on city streets adjacent to highways. Clearflow anticipates traffic patterns, while taking into account sporting/arena events, time of day and weather conditions, and then reflects the back ups and their consequential spill over onto city streets. Often, ClearFlow found it may be faster to stay on the highway instead of seeking alternative side street routes, which involve traffic lights and congestion as well.

Sharing and Embedding Maps

Bing Maps allows users to share maps and embed maps into their websites. By clicking the e-mail icon in the bottom-left corner of Bing Maps, a window will open that displays a shareable URL so others can access the map currently being viewed. This window also provides HTML code to embed a small version of the map onto any web page.

Design

In August 2010, Bing Maps launched an overhauled design for its default view. The new colors create a more visually appealing backdrop for information delivery that helps content 'pop' on the map. The backdrop provides clear differentiation for pushpins, labels and red, yellow and green traffic overlays. These design principles also works well in black and white and creates differentiation for those with the most common forms of color blindness. Also, larger fonts correspond to larger roads to help customers more easily identify main roads in cities. More readable labels eliminate the need for bolding and less-attractive glows. The inclusion of neighborhood labels allows users to quickly find or convey locations in a commonly used and highly relevant format.

Other Features

People, Business and Location Search

The search box at the top of Bing Maps can be used to locate places, businesses and landmarks, and people. Search results appear both on a left-side rail and as pushpins on the map (linked together by numbers). Search results often include addresses, contact information, and reviews for businesses and landmarks. For relevant searches, the user will also see a description of the landmark or place (powered by Wikipedia) if a Wikipedia article exists. The search process can also be guided using local directories for numerous categories (restaurants, hotels, tourist attractions, retail stores, etc).

User Contributions

Bing Maps users can also view and add "user contributed" entries to the map. These user-contributions must be toggled on by users. Such items can include businesses, landmarks, buildings, and locations. Users can browse user-contributions by tags and subscribe to RSS feeds to receive updates of new user-contributions to a specific area.

Dynamic Labels

In August 2010, Bing Maps added dynamic labels to its Silverlight experience (bing.com/maps/explore). Turn on the dynamic labels beta from the map style selector on bing.com/maps/explore and the labels become clickable. This allows users to quickly zoom down to a region or location anywhere on the map with just a few clicks. Zooming back out in a single click is also possible by using the 'breadcrumb' trail at the top left of the map.

AJAX and Silverlight Versions

Bing Maps has two separate versions for users: an AJAX version (located at Bing.com/Maps) and an opt-in Silverlight version (located at Bing.com/Maps/Explore—not available anymore) that requires Microsoft Silverlight to be installed. The Silverlight version is positioned to offer richer, more dynamic features and a smoother experience. In November 2010, the AJAX and Silverlight versions were combined into a semi-hybrid site where Silverlight features such as Map Apps and Streetside could be enabled through the Bing.com/Maps site - these features still required Silverlight to be installed, but does not require use of a separate Bing Maps site.

The AJAX and Silverlight site share the following features: Road View, Aerial View, Bird's-Eye View, Sharing Maps, People/Business/Location Search, Building Footprints, Driving Directions, Walking Directions.

Silverlight users exclusively can use Map Apps, StreetSide View, Photosynths, and Dynamic Labels.

Compatibility

Microsoft states that Bing Maps needs the following environment:

- Windows XP with SP2 or a later version,
- Microsoft. NET Framework 2.0,
- Windows Imaging Component,
- 250 MB or more of hard disk space,
- A 1.0-gigahertz (GHz) processor (2.8 GHz or faster is recommended),
- 256 MB of system memory (1 GB is recommended),
- A 32-MB video card (256 MB is recommended) that supports Microsoft DirectX 9, with 3D hardware acceleration enabled,
- A high-speed or broadband Internet connection.

Compatible browsers include Windows Internet Explorer 6 or later, Mozilla Firefox 3.0 or later, or Safari 3.1 or later. Opera is stated to be usable "with some functionality limitations". Users of browsers that are not considered compatible, as well as users of versions of compatible browsers that are not supported, will be directed away from viewing the map without an error message.

The 3D Maps viewer plug-in requires Microsoft Windows XP Service Pack 2, Microsoft Windows Server 2003, Windows Vista, or Windows 7 with Internet Explorer 6/7/8 or Firefox 1.5/2.0/3.0.

Google Maps

Google Maps is a web mapping service developed by Google. It offers satellite imagery, aerial photography, street maps, 360° panoramic views of streets (Street View), real-time traffic conditions, and route planning for traveling by foot, car, bicycle and air (in beta), or public transportation.

Google Maps began as a C++ desktop program at Where 2 Technologies. In October 2004, the company was acquired by Google, which converted it into a web application. After additional acquisitions of a geospatial data visualization company and a realtime traffic analyzer, Google Maps was launched in February 2005. The service's front end utilizes JavaScript, XML, and Ajax. Google Maps offers an API that allows maps to be embedded on third-party websites, and offers a locator for businesses and other organizations in numerous countries around the world. Google Map Maker allowed users to collaboratively expand and update the service's mapping worldwide but was discontinued from March 2017. However, crowdsourced contributions to Google Maps were not discontinued as the company announced those features will be transferred to the Google Local Guides program.

Google Maps' satellite view is a "top-down" or "birds eye" view; most of the high-resolution imagery of cities is aerial photography taken from aircraft flying at 800 to 1,500 feet (240 to 460 m), while most other imagery is from satellites. Much of the available satellite imagery is no more than three years old and is updated on a regular basis. Google Maps used a variant of the Mercator projection, and therefore could not accurately show areas around the poles. However, in August 2018, the desktop version of Google Maps was updated to show a 3D globe as the default projection. It is still possible to switch back to the Mercator Projection in the settings.

Google Maps for Android and iOS devices was released in September 2008 and features GPS turn-by-turn navigation along with dedicated parking assistance features. In August 2013, it was determined to be the world's most popular app for smartphones, with over 54% of global smartphone owners using it at least once.

In 2012, Google reported having over 7,100 employees and contractors directly working in mapping.

In 2017, along with several other services including YouTube, Chrome, Gmail, Search, and Google Play, Google Maps reached over 1 billion users.

Functionality

Directions and Transit

Google Maps provides a route planner, allowing users to find available directions through driving, public transportation, walking, or biking. Google has partnered globally with over 800 public transportation providers to adopt General Transit Feed Specification (GTFS), making the data available to third parties.

Traffic Conditions

In 2007, Google began offering traffic data as a colored overlay on top of roads and motorways to represent the speed of traffic. Crowdsourcing is used to obtain the GPS-determined locations of a large number of cellphone users, from which live traffic maps are produced.

Google has stated that the speed and location information it collects to calculate traffic conditions is anonymous. Options available in each phone's settings allow users not to share information about their location with Google Maps. Google stated, "Once you disable or opt out of My Location, Maps will not continue to send radio information back to Google servers to determine your handset's approximate location".

Screenshot of Google Maps with traffic option enabled.

Street View

Google Maps car.

On May 25, 2007, Google released Google Street View, a new feature of Google Maps which provides 360° panoramic street-level views of various locations. On the date of release, the feature only included five cities in the US. It has since expanded to thousands of locations around the world. In July 2009, Google began mapping college campuses and surrounding paths and trails.

Street View garnered much controversy after its release because of privacy concerns about the uncensored nature of the panoramic photographs, although the views are only taken on public streets. Since then, Google has begun blurring faces and license plates through automated facial recognition.

In late 2014, Google launched Google Underwater Street View, including 2,300 kilometres (1,400 mi) of the Australian Great Barrier Reef in 3D. The images are taken by special cameras which turn 360 degrees and take shots every 3 seconds.

In 2017, in both Google Maps and Google Earth, Street View navigation of the International Space Station interior spaces became available.

45° Imagery

An example of the Leaning Tower of Pisa in the 45° view.

In December 2009, Google introduced a new view consisting of 45° angle aerial imagery, offering a "bird's eye view" of cities. The first cities available were San Jose and San Diego. This feature was initially available only to developers via the Google Maps API. In February 2010, it was introduced as an experimental feature in Google Maps Labs. In July 2010, 45° imagery was made available in Google Maps in select cities in the United States and worldwide.

Business Listings

Google collates business listings from multiple on-line and off-line sources. To reduce duplication in the index, Google's algorithm combines listings automatically based on address, phone number, or geocode, but sometimes information for separate businesses will be inadvertently merged with each other, resulting in listings inaccurately incorporating elements from multiple businesses. Google allows business owners to verify their own business data through Google My Business, and has also recruited volunteers to check and correct ground truth data.

Google Maps can easily be manipulated by businesses which aren't physically located in the area they record a listing. There are cases of people abusing Google Maps to overtake their competition where they place a number of unverified listings on online directory sites knowing the information will roll across to Google (duplicate sites). The people that update these listings do not use a registered business name. Keywords and location details are placed on their Google Maps business title which overtake credible business listings. In Australia in particular, genuine companies and businesses are noticing a trend of fake business listings in a variety of industries.

Indoor Maps

In March 2011, indoor maps were added to Google Maps, giving users the ability to navigate themselves within buildings such as airports, museums, shopping malls, big-box stores, universities, transit stations, and other public spaces (including underground facilities). Google encourages owners of public facilities to submit floor plans of their buildings in order to add them to the service. Map users can view different floors of a building or subway station by clicking on a level selector that is displayed near any structures which are mapped on multiple levels.

My Maps

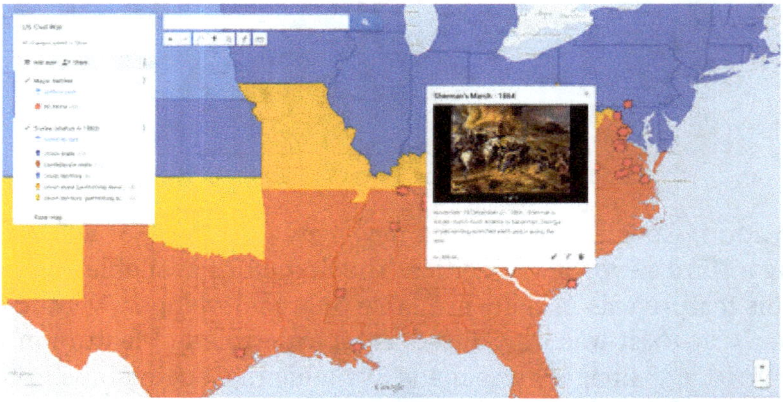

Google My Maps.

My Maps is a feature in Google Maps launched in April 2007 that enables users to create custom maps for personal use or sharing. Users can add points, lines, shapes, notes and images on top of Google Maps using a WYSIWYG editor. An Android app for My Maps, initially released in March 2013 under the name Google Maps Engine Lite, is also available.

Google Local Guides

Google Local Guides is a program launched by Google Maps to enable its users to contribute to Google Maps and provide them additional perks and benefits for the work. The program is partially a successor to Google Map Maker as features from the former program became integrated into the website and app.

The program consists of adding reviews, photos, basic information, videos and correcting information such as wheelchair accessibility.

Dark Mode

The maps app will also have a theme for navigation mode for users in line with current trends.

Implementation

As the user drags the map, the grid squares are downloaded from the server and inserted into the page. When a user searches for a business, the results are downloaded in the background for insertion into the side panel and map; the page is not reloaded. A hidden iframe with form submission is used because it preserves browser history. Like many other Google web applications, Google Maps uses JavaScript extensively. The site also uses JSON for data transfer rather than XML, for performance reasons.

A split view screenshot of Google Maps. In the bottom half the *Street Maps* is shown, while in the top half the *Street View* is shown. A user can zoom-in and out either of them independently of the zoom level of each.

The version of Google Street View for classic Google Maps required Adobe Flash. In October 2011, Google announced Maps GL, a WebGL version of Maps with better renderings and smoother transitions. Indoor maps uses JPG, .PNG, .PDF, .BMP, or GIF, for floor plans.

Users who are logged into a Google Account can save locations so that they are overlaid on the map with various colored "pins" whenever they browse the application. These "Saved places" can be organised into user named lists and shared with other users. One default list "Starred places" also automatically creates a record in another Google product, Google Bookmarks.

Map Data and Imagery

The Google Maps terms and conditions state that usage of material from Google Maps is regulated by Google Terms of Service and some additional restrictions. Google has either purchased local map data from established companies, or has entered into lease agreements to use copyrighted map data. The owner of the copyright is listed at the bottom of zoomed maps. For example, street maps in Japan are leased from Zenrin. Street maps in China are leased from Auto Navi. Russian street maps are leased from Geocentre Consulting and Tele Atlas. Data for North Korea is sourced from the companion project Google Map Maker.

Street map overlays, in some areas, may not match up precisely with the corresponding satellite images. The street data may be entirely erroneous, or simply out of date: "The biggest challenge is the currency of data, the authenticity of data," said Google Earth representative Brian McClendon. As a result, in March 2008 Google added a feature to edit the locations of houses and businesses.

Restrictions have been placed on Google Maps through the apparent censoring of locations deemed potential security threats. In some cases the area of redaction is for specific buildings, but in other cases, such as Washington, D.C., the restriction is to use out dated imagery.

Google Maps API

After the success of reverse-engineered mashups such as chicagocrime.org and housingmaps.com, Google launched the Google Maps API in June 2005 to allow developers to integrate Google Maps into their websites. It was a free service that didn't require an API key until June 2018 (changes went into effect on July 16), when it was announced that an API key linked to a Google Cloud account with billing enabled would be required to access the API. The API currently does not contain ads, but Google states in their terms of use that they reserve the right to display ads in the future.

By using the Google Maps API, it is possible to embed Google Maps into an external website, on to which site-specific data can be overlaid. Although initially only a JavaScript API, the Maps API was expanded to include an API for Adobe Flash applications (but this has been deprecated), a service for retrieving static map images, and web services for performing geocoding, generating driving directions, and obtaining elevation profiles. Over 1,000,000 web sites use the Google Maps API, making it the most heavily used web application development API. In September 2011, Google announced it would deprecate the Google Maps API for Flash.

The Google Maps API is free for commercial use, provided that the site on which it is being used is publicly accessible and does not charge for access, and is not generating more than 25,000 map accesses a day. Sites that do not meet these requirements can purchase the Google Maps API for Business.

As of 21 June 2018, Google increased the prices of the Maps API and requires a billing profile.

Google Maps in China

Due to restrictions on geographic data in China, Google Maps must partner with a Chinese digital map provider in order to legally show Chinese map data. Since 2006, this partner has been AutoNavi.

Within China, the State Council mandates that all maps of China use the GCJ-02 coordinate system, which is offset from the WGS-84 system used in most of the world. google.*cn*/maps (formerly Google Ditu) uses the GCJ-02 system for both its street maps and satellite imagery. google.*com*/maps also uses GCJ-02 data for the street map, but uses WGS-84 coordinates for satellite imagery, causing the so-called China GPS shift problem.

Frontier alignments also present some differences between google.*cn*/maps and google.*com*/maps. On the latter, sections of the Chinese border with India and Pakistan are shown with dotted lines, indicating areas or frontiers in dispute. However, google.*cn* shows the Chinese frontier strictly according to Chinese claims with no dotted lines indicating the border with India and Pakistan. For example, the South Tibet region claimed by China but administered by India as a large part of Arunachal Pradesh is shown inside the Chinese frontier by google.*cn*, with Indian highways ending abruptly at the Chinese claim line. Google.*cn* also shows Taiwan and the South China Sea Islands as part of China. Google Ditu's street map coverage of Taiwan no longer omits major state organs, such as the Presidential Palace, the five Yuans, and the Supreme Court.

Feature-wise, google.*cn*/maps does not feature My Maps. On the other hand, while google.*cn* displays virtually all text in Chinese, google.*com*/maps displays most text (user-selectable real text as well as those on map) in English. This behavior of displaying English text is not consistent but intermittent – sometimes it is in English, sometimes it is in Chinese. The criteria for choosing which language is displayed are not known publicly.

Discontinued Features

Google Latitude

Google Latitude was a feature from Google that lets users share their physical locations with other people. This service was based on Google Maps, specifically on mobile devices. There was an iGoogle widget for desktops and laptops as well. Some concerns were expressed about the privacy issues raised by the use of the service. On August 9, 2013, this service was discontinued, and in March 22, 2017, Google incorporated the features from Latitude into the Google Maps app.

Google Map Maker

In areas where Google Map Maker was available, for example, much of Asia, Africa, Latin America and Europe as well as the United States and Canada, anyone who logged into their Google account could directly improve the map by fixing incorrect driving directions, adding biking trails, or adding a missing building or road. General map errors in Australia, Austria, Belgium, Denmark, France, Liechtenstein, Netherlands, New Zealand, Norway, South Africa, Switzerland, and the United States could be reported using the Report a Problem link in Google Maps and would be updated by Google. For areas where Google used Tele Atlas data, map errors could be reported using Tele Atlas map insight.

If imagery was missing, out dated, misaligned, or generally incorrect, one could notify Google through their contact request form.

In November 2016, Google announced the discontinuation of Google Map Maker as of March 2017.

Mobile App

Google Maps is available as a mobile app for the Android and iOS mobile operating systems. The Android app was first released in September 2008, though the GPS-localization feature had been in testing on cellphones since 2007. Up until iOS 6, the built-in maps application on the iOS operating system was powered by Google Maps. However, with the announcement of iOS 6 in June 2012, Apple announced that they had created their own Apple Maps mapping service, which officially replaced Google Maps when iOS 6 was released on September 19, 2012. However, at launch, Apple Maps received significant criticism from users due to inaccuracies, errors and bugs. One day later, *The Guardian* reported that Google was preparing its own Google Maps app, which was released on December 12, 2012. Within only two days, the application had been downloaded over ten million times.

Features

The Google Maps apps for iOS and Android have many of the same features, including turn-by-turn navigation, street view, and public transit information. Turn-by-turn navigation was originally announced by Google as a separate beta testing app exclusive to Android 2.0 devices in October 2009. The original standalone iOS version did not support the iPad, but tablet support was added with version 2.0 in July 2013. An update in June 2012 for Android devices added support for offline access to downloaded maps of certain regions, a feature that was eventually released for iOS devices, and made more robust on Android, in May 2014.

At the end of 2015 Google Maps announced its new offline functionality, but with various limitations – downloaded area cannot exceed 120,000 square kilometres and require considerable amount of storage space. In January 2017, Google added a feature exclusively to Android that will, in some U.S. cities, indicate the level of difficulty in finding available parking spots, and on both Android and iOS, the app can, as of an April 2017 update, remember where users parked. In August 2017, Google Maps for Android was updated with new functionality to actively help the user in finding parking lots and garages close to a destination. In December 2017, Google added a new two-wheeler mode to its Android app, designed for users in India, allowing for more accessibility in traffic conditions.

Reception

USA Today welcomed the application back to iOS, saying: "The reemergence in the middle of the night of a Google Maps app for the iPhone is like the return of an old friend. Only your friend, who'd gone missing for three months, comes back looking better than ever." Jason Parker of *CNET*, calling it "the king of maps", said, "With its iOS Maps app, Google sets the standard for what mobile navigation should be and more." Bree Fowler of the *Associated Press* compared Google's and Apple's map applications, saying: "The one clear advantage that Apple has is style. Like Apple devices, the maps are clean and clear and have a fun, pretty element to them, especially in 3-D. But when it comes down to depth and information, Google still reigns superior and will no doubt be welcomed back by its fans." *Gizmodo* gave it a ranking of 4.5 stars, stating: "Maps Done Right". According to *The New York Times*, Google "admits that it's [iOS app is] even better than Google Maps for Android phones, which has accommodated its evolving feature set mainly by piling on menus".

However, Google Maps' location tracking is widely regarded as a threat to users' privacy, with Dylan Tweney of *VentureBeat* writing in August 2014 that "Google is probably logging your location, step by step, via Google Maps", and linked users to Google's location history map, which "lets you see the path you've traced for any given day that your smartphone has been running Google Maps". Tweney then provided instructions on how disable location history. The history tracking was also noticed, and recommended disabled, by editors at *CNET* and *TechCrunch*. Additionally, *Quartz* reported in April 2014 that a "sneaky new privacy change" would have an effect on the majority of iOS users. The privacy change, an update to the Gmail iOS app that "now supports sign-in across Google iOS apps, including Maps, Drive, YouTube and Chrome", meant that Google would be able to identify users' actions across its different apps.

Potential Misuse

In 2005 the Australian Nuclear Science and Technology Organization (ANSTO) complained about the potential for terrorists to use the satellite images in planning attacks, with specific reference to the Lucas Heights nuclear reactor; however, the Australian Federal government did not support the organization's concern. At the time of the ANSTO complaint, Google had colored over some areas for security (mostly in the US), such as the rooftop of the White House and several other Washington, D.C., US buildings.

In October 2010, Nicaraguan military commander Edén Pastora stationed Nicaraguan troops on the Isla Calero (in the delta of the San Juan River), justifying his action on the border delineation given by Google Maps. Google has since updated its data which it found to be incorrect.

On January 27, 2014, documents leaked by Edward Snowden revealed that the NSA and the GCHQ intercepted Google Maps queries made on smartphones, and used them to locate the users making these queries. One leaked document, dating to 2008, stated that "it effectively means that anyone using Google Maps on a smartphone is working in support of a GCHQ system."

References

- "New Bird's Eye Imagery & Streetside Coverage on Bing Maps!". Bing Maps Blog. Microsoft. May 8, 2012. Archived from the original on 12 August 2012. Retrieved 9 August 2012
- Goodchild, M.F. (2007). "Citizens as sensors: the world of volunteered geography". Geojournal. 69 (4): 211–221. Citeseerx 10.1.1.525.2435. Doi:10.1007/s10708-007-9111-y
- Butler, Patrick (2014-04-10). "Collaborative mapping | Collaborative mapping". Espatial.com. Retrieved 2016-01-15
- What-is-a-web-map, web, guide: axismaps.com, Retrieved 14 February, 2019
- Perez, Sarah. "Google to shut down Map Maker, its crowdsourced map editing tool | techcrunch". Archived from the original on August 11, 2017. Retrieved June 23, 2017

5

Interdisciplinary Fields Involving Cartography

There are numerous fields which involve the application of cartography. A few of them are geodesy, geomatics and topography. Geodesy deals with the study of the shape and area of the Earth, geomatics focuses on the collection, storage and analysis of geographic data or geographic information and topography studies the shape and features of land surfaces. The diverse applications of cartography in these fields have been thoroughly discussed in this chapter.

GEODESY

Geodesy is the science of accurately measuring the Earth's size, shape, orientation, mass distribution and how these vary with time.

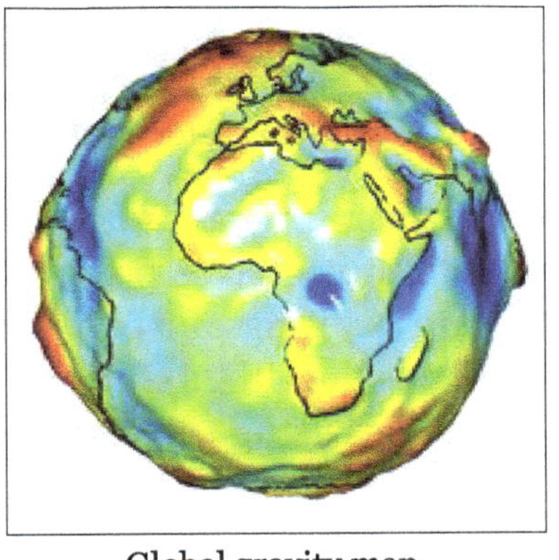

Global gravity map.

Over the last century, geodesy has developed from fairly simple surveying technologies, which helped to accurately determine positions on Earth, to a complex toolbox of methods now available to scientific researchers and students. In recent decades, geodetic applications have rapidly expanded from measuring plate motions and monitoring earthquake hazards to include research on volcanic, landslide, and weather hazards; climate change; and water resources.

Geodetic Methods

GPS/GNSS (Global positioning system/Global navigation satellite system)

High precision GPS station in the Sierra Nevada region of in the Plate Boundary Observatory (station P149).

Three GPS satellites used for positioning.

Three GPS satellites are used for positioning and a fourth provides a time correction. Together they allow calculation of precise positions. GPS is a USA-based fleet of more than 30 satellites that are orbiting our planet approximately 11,000 miles above Earth's surface. GNSS includes the USA's GPS and similar satellites from other countries. A position can be calculated using three satellites, plus a fourth to correct for clock imprecision. You may already be familiar with handheld GPS units, which people have in their phones, tablets, cameras, cars, and more. Whereas handheld GPS can be accurate to a few meters or tens-of-meters, the high-precision "differential" GPS units that Earth scientists use in their research can measure motions as slow as one millimeter per year. The first major applications of high-precision GPS were to monitor plate tectonic motions and assess earthquake and volcanic hazards. More recently, scientists have been able to apply the technique to landslide hazards, ground water monitoring, tide gauging, ice/snow monitoring, and soil and atmospheric moisture.

Lidar (Light Detection and Ranging)

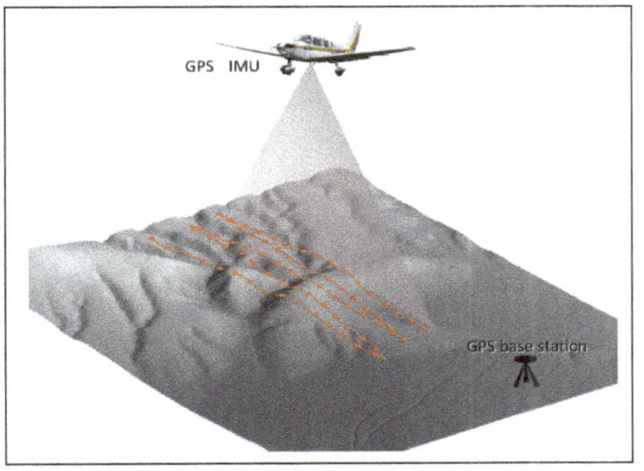

With airborne LiDAR a scanner is mounted on an airplane and combined with data from GPS and an IMU (inertial measurement unit) to yield high-resolution topographic data.

Lidar is a remote sensing technology that measures distance by sending out laser pulses and calculating the reflection return time. Lidar scanners can be mounted on aircraft, ground-based tripods, or mobile devices (airborne lidar, terrestrial laser scanning [TLS], and mobile lidar, respectively). Depending on the way a survey is set up, the resulting topographic model can have a resolution of meters to centimeters. The laser beams also have the ability to penetrate and return through holes in a vegetation canopy, thus yielding a "bare earth" topography from the latest returning signals that is not possible using other methods. Differences between first and last returns in vegetated areas can yield canopy volume and density. Repeated scans of the same area allow for detailed measurement of topographic change. Lidar can be used for a broad range of hazard assessment applications, stratigraphic analyses, understanding geomorphic and tectonic processes, and vegetation studies.

InSAR (Interferometric Synthetic Aperture Radar)

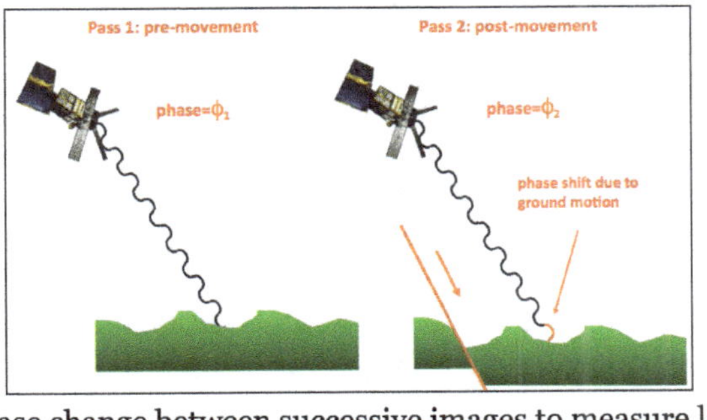

InSAR uses the phase change between successive images to measure land level changes. This example shows the method applied to measuring change caused by an earthquake.

InSAR measures ground deformation using two of more synthetic aperture radar (SAR) images. Most commonly, the images are from Earth-orbiting radar satellites but the method can be used from aircraft or ground-based sensors too. The radar signal phase changes between repeat images allows for centimeter-scale measurement of deformation over spans of days to years and over large regions. Although complications can arise from ground-surface moisture and changing atmospheric conditions, radar is able penetrate clouds and provide data over large areas, thus making it a good complement to other methods such as GPS, lidar, and SfM which have more limited spatial extents. InSAR has applications for monitoring of natural hazards (ex. earthquakes, volcanoes and landslides), measuring land subsidence, and even surface water level assessments and glacial ice velocity.

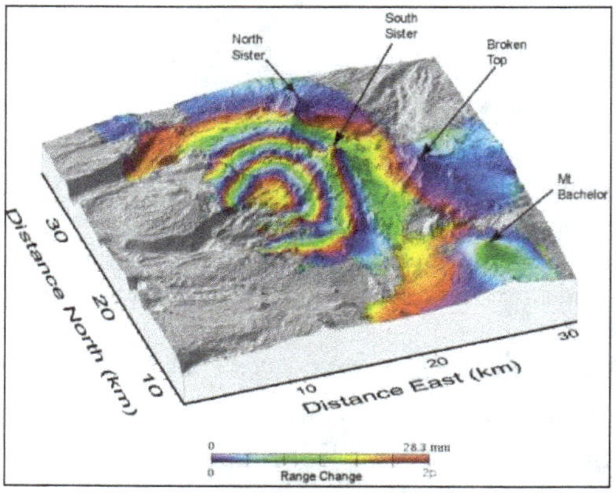

Interferogram showing volcanic uplift about
3 miles west of South Sister, Oregon.

Gravity Recovery and Climate Experiment and Follow-on Mission

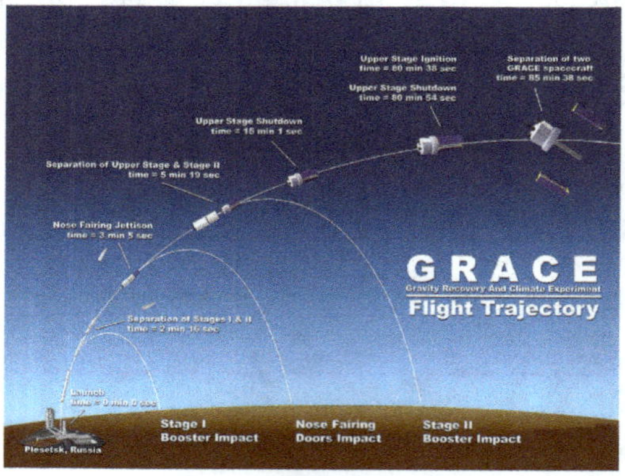

The flight trajectory of the GRACE twin satellite system is shown.

Measuring Earth's gravity field, is also an element of geodesy. The advent of satellite-based gravity measurements has profoundly affected our ability to determine the changing mass distribution on Earth. GRACE (Gravity Recovery And Climate Experiment) has led to unprecedented observations. The gravity field of the Earth is uneven, reflecting the mass distribution on our planet. The orbit of the twin satellites of GRACE is disturbed by the uneven gravity field, changing the distance between the satellites. This distance change is measured using a microwave ranging system. This technique is used in tandem with GPS, as each of the satellites is equipped with a highly precise GPS receiver. This measure of the gravity of Earth can be used for many applications, but changes in groundwater and ice mass have been two of the most profound. They have aided researchers in understanding climate change impacts and groundwater changes over time. GRACE data can be used to track the distribution of water over Earth's surface on continents, ice sheet volume, sea level change, ocean currents, and the dynamics of the inner structure of Earth.

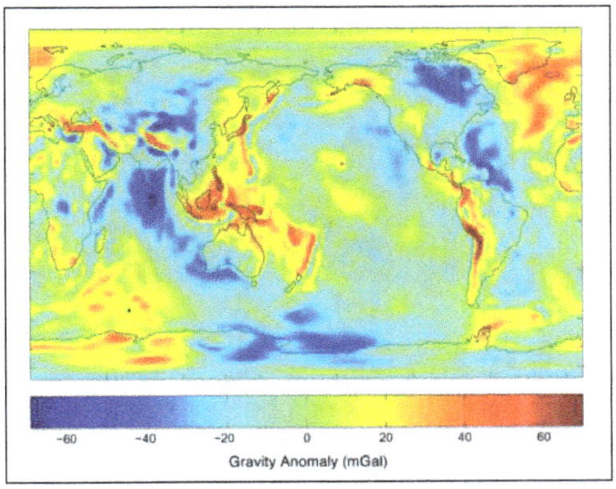

One of the first produced gravity maps of Earth from GRACE, based on 111 days of data in 2003.

Altimetry: Ice and sea Level

Satellite altimetry measures the distance between a satellite and a target on Earth. Usually, this is done with a radar altimetry system, which sends a radar pulse to Earth's surface and then measures the time the pulse takes to reach the surface and return to estimate the distance. Particular characteristics of the signal, like the magnitude and shape of the waveform, give information on the type of surface surveyed. Other altimetry systems exist, like ATLAS (Advanced Topographic Laser Altimeter System), a laser altimetry system on ICESat-2.

These methods are used to survey both sea level and ice altitude. These satellite missions last for years, so collect data ideal for climate change studies as the ice and sea levels can be measured over time. This data can be compared to data collected from GRACE, to give a complete picture of how ice volume and sea level are changing.

Structure-from-Motion (SfM) Photogrammetry

Structure from Motion or SfM is a photogrammetric method for creating three-dimensional models of a feature or topography from overlapping two-dimensional photographs taken from many locations and orientations to reconstruct the photographed scene. The applications of SfM are wide-ranging, from many sub-fields of geoscience (geomorphology, tectonics, structural geology, geodesy, and mining) to archaeology, architecture, and agriculture. In addition to ortho-rectified imagery, SfM produces a dense point cloud data set that is similar in many ways to that produced by airborne or terrestrial lidar. The advantages of SfM are its relative cost in comparison to lidar, as well as its ease of use. The only required equipment is a camera.

A computer and software are needed for data processing. Additionally, an aerial platform like a balloon or drone can also be useful for topographic mapping applications. Because SfM relies on optical imagery, it is not able to generate the "bare earth" topographic products that are typical derivatives of lidar-based technologies – thus, SfM is usually best suited to areas of limited vegetation.

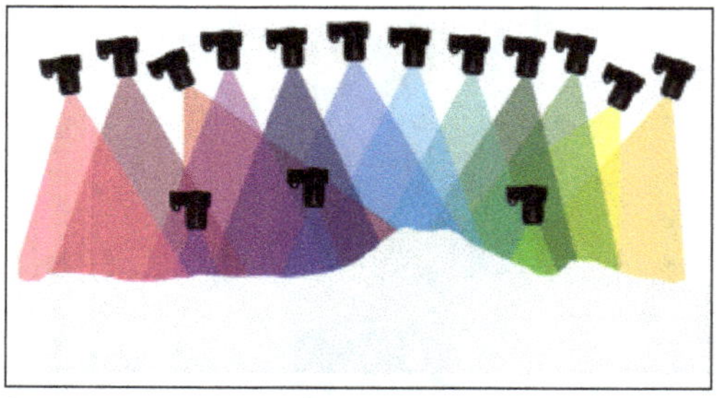

A cartoon of the SfM technique, based on taking photographs from a wide array of orientations and distances. The location of the camera for each photograph is calculated using features recognized in multiple photographs.

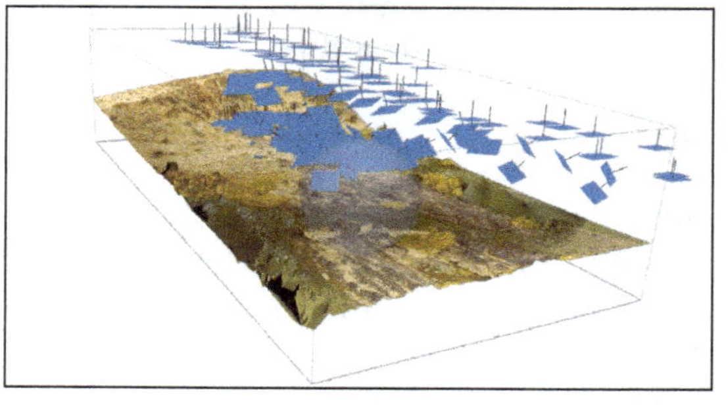

An example SfM model from the Pofadder shear zone. The blue rectangles indicate the calculated camera locations; the model is a 3D point cloud with a photo overlay.

Meters: Borehole, Tilt and Creep

Three types of meters can supplement the data collected using the geodetic techniques detailed above: Borehole strainmeters, tiltmeters and creepmeters.

Borehole strainmeters are installed in boreholes and measure very small changes in the dimension of a borehole at depth, reflecting the continuous strain in Earth's crust. This is accomplished by measuring the change in diameter or volume of a strainmeter installed in the borehole. Usually, strainmeters are installed at a depth of 200 m in a 15 cm diameter borehole. A seismometer is installed above the strainmeter. A tiltmeter may also be installed in the borehole.

Tiltmeters are very sensitive inclinometers that measure the change from horizontal. These may be installed in boreholes with borehole strainmeters. A tiltmeter may also be installed on the ground surface. Tiltmeters are generally used for fault monitoring, volcano monitoring, dam monitoring, assessing potential landslides, and the orientation and volume of hydraulic fractures.

Creepmeters are used exclusively to quantify fault slip. Two monuments are placed on either side of a fault, 30 meters apart. A wire connects the two monuments, and the displacement of the wire represents the displacement on the fault.

GEOMATICS

A surveyor's shed showing equipment used for geomatics.

Geomatics is defined in the ISO/TC 211 series of standards as the "discipline concerned with the collection, distribution, storage, analysis, processing, presentation of geographic data or geographic information". Under another definition it "consists of products, services and tools involved in the collection, integration and management of geographic data". It includes geomatics engineering (and surveying engineering) and is related to geospatial science (also geospatial engineering and geospatial technology).

Michel Paradis, a French-Canadian surveyor, introduced *geomatics* as a new scientific term in an article published in 1981 in *The Canadian Surveyor* and in a keynote address at the centennial congress of the Canadian Institute of Surveying in April 1982. He claimed that at the end of the 20th century the needs for geographical information would reach a scope without precedent in history and in order to address these needs, it was necessary to integrate in a new discipline both the traditional disciplines of land surveying and the new tools and techniques of data capture, manipulation, storage and diffusion.

Geomatics includes the tools and techniques used in land surveying, remote sensing, cartography, geographic information systems (GIS), global-navigation satellite systems (GPS, GLONASS, Galileo, Compass), photogrammetry, geophysics, geography, and related forms of earth mapping. The term was originally used in Canada, because it is similar in origin to both French and English, but has since been adopted by the International Organization for Standardization, the Royal Institution of Chartered Surveyors, and many other international authorities, although some (especially in the United States) have shown a preference for the term *geospatial technology*.

The related field of *hydrogeomatics* covers the area associated with surveying work carried out on, above or below the surface of the sea or other areas of water. The older term of hydrographics was considered too specific to the preparation of marine charts, and failed to include the broader concept of positioning or measurements in all marine environments.

A *geospatial network* is a network of collaborating resources for sharing and coordinating geographical data and data tied to geographical references. One example of such a network is the Open Geospatial Consortium's efforts to provide *ready global access to geographic information.*

A number of university departments which were once titled "surveying", "survey engineering" or "topographic science" have re-titled themselves using the terms "geomatics" or "geomatic engineering".

The rapid progress and increased visibility of geomatics since the 1990s has been made possible by advances in computer hardware, computer science, and software engineering, as well as by airborne and space observation remote-sensing technologies.

The science of deriving information about an object using a sensor without physically contacting it is called remote sensing, which is a part of geomatics.

Science

Geospatial science is an academic discipline incorporating fields such as surveying, geographic information systems, hydrography and cartography. Spatial science is typically concerned with the measurement, management, analysis and display of spatial information describing the Earth, its physical features and the built environment.

The term spatial science or spatial sciences is primarily used in Australia. Australian universities which offer degrees in spatial science include Curtin University, the University of Tasmania, the University of Adelaide,Melbourne University and RMIT University.

In the U.S., Texas A&M University offers a bachelor's degree in Spatial Sciences and is home to its own Spatial Sciences Laboratory. Beginning in 2012, the University of Southern California started to place more emphasis on the spatial science branch of its geography department, with traditional human and physical geography courses and concentrations either not being offered on a regular basis or phased out. In place, the university now offers graduate programs strictly related to spatial science and its geography department offers a spatial science minor rather than the original geography major.

Spatial information practitioners within the Asia-Pacific region are represented by the professional body called the Surveying and Spatial Sciences Institute (SSSI).

Engineering

Geomatics Engineering, Geomatic Engineering, Geospatial Engineering is a rapidly developing engineering discipline that focuses on spatial information (i.e. information that has a location). The location is the primary factor used to integrate a very wide range of data for spatial analysis and visualization. Geomatics engineers apply engineering principles to spatial information and implement relational data structures involving measurement sciences, thus using geomatics and acting as spatial information engineers. Geomatics engineers manage local, regional, national and global spatial data infrastructures. Geomatics Engineering also involves aspects of Computer Engineering, Software Engineering and Civil Engineering.

Geomatics is a field that incorporates several others such as the older field of land surveying engineering along with many other aspects of spatial data management ranging from data science and cartography to geography. Following the advanced developments in digital data processing, the nature of the tasks required of the professional land surveyor has evolved and the term "surveying" no longer accurately covers the whole range of tasks that the profession deals with. As our societies become more complex, information with a spatial position associated with it becomes more critical to decision-making, both from a personal and a business perspective, and also from a community and a large-scale governmental viewpoint.

Therefore, the geomatics engineer can be involved in an extremely wide variety of information gathering activities and applications. Geomatics engineers design, develop, and operate systems for collecting and analyzing spatial information about the land, the oceans, natural resources, and manmade features.

The more traditional land surveying strand of geomatics engineering is concerned with the determination and recording of boundaries and areas of real property parcels, and the preparation and interpretation of legal land descriptions. The tasks more closely related to civil engineering include the design and layout of public infrastructure and urban subdivisions, and mapping and control surveys for construction projects.

Geomatics engineers serve society by collecting, monitoring, archiving, and maintaining diverse spatial data infrastructures. Geomatics engineers utilize a wide range of technologically advanced tools such as digital theodolite/distance meter total stations, Global Positioning System (GPS) equipment, digital aerial imagery (both satellite and air-borne), and computer-based geographic information systems (GIS). These tools enable the geomatics engineer to gather, process, analyze, visualize and manage spatially related information to solve a wide range of technical and societal problems.

Geomatics engineering is the field of activity that integrates the acquisition, processing, analysis, display and management of spatial information. It is an exciting and new grouping of subjects in the spatial and environmental information sciences with a broad range of employment opportunities as well as offering challenging pure and applied research problems in a vast range of interdisciplinary fields.

In different schools and in different countries the same education curriculum is administered with the name surveying in some, and in others with the names geomatics, civil engineering surveying, geomatics engineering, geospatial (information) engineering, surveying engineering, or geodesy and geoinformatics. While these occupations were at one time often taught in civil engineering education programs, more and more universities include the departments relevant for geo-data sciences under informatics, computer science or applied mathematics. These facts demonstrate the breadth, depth and scope of the highly interdisciplinary nature of geomatics engineering. The job of *geospatial engineer* is well established in the U.S. military.

Applications

Application areas include:

- Aeromagnetic surveys,
- Airborne geophysics,
- Air navigation services,
- Archaeological excavation and survey for GIS applications,
- Coastal zone management and mapping,
- Disaster informatics for disaster risk reduction and response,
- The environment,
- Infrastructure management,
- Land management and reform,
- Natural resource monitoring and development,
- Seismic Interpretation,
- Subdivision planning,
- Urban and regional planning,
- Oceanography,
- Meteorology,
- Parks,
- Resource Management,
- Climate Change/Environmental Monitoring.

TOPOGRAPHY

Topography is a broad term used to describe the detailed study of the earth's surface. This includes changes in the surface such as mountains and valleys as well as features such as rivers and roads. It can also include the surface of other planets, the moon, asteroids and meteors. Topography is closely linked to the practice of surveying, which is the practice of determining and recording the position of points in relation to one another.

Some of the first known topographic surveys were conducted by the British military in the late eighteenth century. In the United States, the earliest detailed surveys were made during the War of 1812 by the "Topographical Bureau of the Army." Throughout the twentieth century topographical mapping became more complex and precise with the invention of instruments such as theodolites and automatic levels. Most recently, developments in the digital world such as GIS (geographic information system) have allowed us to create increasingly complex topographical maps.

Objectives

Modern-day topography is generally concerned with the measurement and recording of elevation contours, producing a three-dimensional representation of the earth's surface. A series of points are chosen and measured in terms of their horizontal coordinates, such as latitude and longitude, and their vertical position, in terms of altitude. When recorded in a series, these points produce contour lines which show gradual changes in the terrain.

Techniques

The most widely used form of measurement is known as Direct Survey. This is the process of manually measuring distances and angles using leveling instruments such as theodolites. Direct surveying provides the basic data for all topographic mapping, including digital imaging systems. This information can be used in conjunction with other systems such as aerial photography or satellite imagery to provide a complete picture of the land in question.

Sonar mapping is the primary technique used to map the ocean floor. A pulse of sound is sent through the water from an underwater speaker, and is reflected back again by objects in the water, such as the ocean bottom, coral beds, or a submarine. Microphones measure the reflected sound waves. The time that the echo takes to return is proportional to the distance of the reflecting object. This data allows changes in the underwater terrain and other objects likes shipwrecks to be mapped.

Applications

A topographic study can be used for a variety of applications such as military planning and geological exploration. Detailed information about terrain and surface features is also essential for the planning and construction of any major civil engineering or construction projects. More recently, large scale surveys such as Google Maps have been produced using satellite technology, providing the first complete, widely available surveys of the earth.

Digital Mapping Systems

There are a variety of digital systems which utilize the basic data collected from topographic surveying to produce maps:

- GIS uses computer software to create highly detailed maps with distinct layers displaying almost any type of element, such as roads, bridges, buildings, rivers, political boundaries, soil types,
- 3-D rendering uses satellite or aerial images to produce a three-dimensional model using computer software,
- Aerial photography and photogrammetry combine photos from different angles and use the process of triangulation to calculate the location of elements.

Permissions

All chapters in this book are published with permission under the Creative Commons Attribution Share Alike License or equivalent. Every chapter published in this book has been scrutinized by our experts. Their significance has been extensively debated. The topics covered herein carry significant information for a comprehensive understanding. They may even be implemented as practical applications or may be referred to as a beginning point for further studies.

We would like to thank the editorial team for lending their expertise to make the book truly unique. They have played a crucial role in the development of this book. Without their invaluable contributions this book wouldn't have been possible. They have made vital efforts to compile up to date information on the varied aspects of this subject to make this book a valuable addition to the collection of many professionals and students.

This book was conceptualized with the vision of imparting up-to-date and integrated information in this field. To ensure the same, a matchless editorial board was set up. Every individual on the board went through rigorous rounds of assessment to prove their worth. After which they invested a large part of their time researching and compiling the most relevant data for our readers.

The editorial board has been involved in producing this book since its inception. They have spent rigorous hours researching and exploring the diverse topics which have resulted in the successful publishing of this book. They have passed on their knowledge of decades through this book. To expedite this challenging task, the publisher supported the team at every step. A small team of assistant editors was also appointed to further simplify the editing procedure and attain best results for the readers.

Apart from the editorial board, the designing team has also invested a significant amount of their time in understanding the subject and creating the most relevant covers. They scrutinized every image to scout for the most suitable representation of the subject and create an appropriate cover for the book.

The publishing team has been an ardent support to the editorial, designing and production team. Their endless efforts to recruit the best for this project, has resulted in the accomplishment of this book. They are a veteran in the field of academics and their pool of knowledge is as vast as their experience in printing. Their expertise and guidance has proved useful at every step. Their uncompromising quality standards have made this book an exceptional effort. Their encouragement from time to time has been an inspiration for everyone.

The publisher and the editorial board hope that this book will prove to be a valuable piece of knowledge for students, practitioners and scholars across the globe.

Index

www.ingramcontent.com/pod-product-compliance
Lightning Source LLC
LaVergne TN
LVHW080608210926
848485LV00003B/122

* 9 7 8 1 6 4 1 1 6 5 7 4 7 *